Collins Primary Maths Pupil Book

Series Editor: Peter Clarke

Authors: Peter Clarke and Jeanette Mumford

Contents

Calculator methods	To carry out calculations with more than one step using brackets and the memory To enter numbers and interpret the display in different contexts (decimals, percentages, money, metric measures)	62–64
Checking results	To check a result by considering whether it is of the right order of magnitude and by working the problem backwards	65
Equations, formulae and identities	To use letter symbols to represent unknown numbers or variables; know the meanings of the words term, expression and equation To understand that algebraic operations follow the same conventions and order as arithmetic operations To simplify linear algebraic expressions by collecting like terms; begin to multiply a single term over a bracket (integer coefficients) To construct and solve simple linear equations with integer coefficients (unknown on one side only) using an appropriate method (e.g. inverse operations) To use simple formulae from mathematics and other subjects; substitute positive integers into simple linear expressions and formulae and, in simple cases, derive a formula	66–71
Sequences, functions and graphs	To generate and describe simple integer sequences To generate terms of a simple sequence, given a rule (e.g. finding a term from the previous term, finding a term given its position in the sequence) To express simple functions in words, then using symbols; represent them in mappings To generate coordinate pairs that satisfy a simple linear rule; plot the graphs of simple linear functions, where y is given explicitly in terms of x, on paper and using ICT; recognise straight-line graphs parallel to the x-axis or y-axis To begin to plot and interpret the graphs of simple linear functions arising from real-life situations	72–81
Geometrical reasoning: lines, angles and shapes	To use the vocabulary, notation and labelling conventions for lines, angles and shapes To identify parallel and perpendicular lines; know the sum of angles at a point, on a straight line and in a triangle, recognise vertically opposite angles To begin to identify and use angle, side and symmetry properties of triangles and quadrilaterals; solve geometrical problems involving these properties To use 2-D representations to visualise 3-D shapes and deduce some of their properties	82–88
Transformations	To understand and use the language and notation associated with reflections, translations and rotations To recognise and visualise the transformation and symmetry of a 2-D shape: – reflection in given mirror lines, and line symmetry; rotation about a given point and rotation symmetry; to explore these transformations and symmetries using ICT	89–96
Co-ordinates	To use conventions and notation for 2-D co-ordinates in all four quadrants; find co-ordinates of points determined by geometric information	97–98
Construction	To use a ruler and protractor to: – measure and draw lines to the nearest millimetre and angles, including reflex angles, to the nearest degree; – construct a triangle given two sides and the included angle (SAS); or two angles and the included side (ASA); – explore these constructions using ICT To use ruler and protractor to construct simple nets of 3-D shapes	99–103
Measures and mensuration	To use names and abbreviations of units of measurement to measure, estimate, calculate and solve problems in everyday contexts including length, area, mass, capacity, time and angle; convert one metric unit to another; read and interpret scales on a range of measuring instruments To use an angle measure; distinguish between and estimate the size of acute, obtuse and reflex angles To know and use the formula for the area of a rectangle; calculate the perimeter and area of shapes made from rectangles To calculate the surface area of cubes and cuboids	104–113
Specifying a problem, planning and collecting data	To decide which data would be relevant to an enquiry and possible sources To plan how to collect and organise small sets of data; design a data collection sheet or questionnaire to use in a simple survey; construct frequency tables for discrete data, grouped where appropriate in equal class intervals To collect small sets of data from surveys and experiments, as planned	114–115
Processing and representing data, using ICT as appropriate	To calculate statistics for small sets of discrete data: - find the mode and range and the modal class for grouped data To construct, on paper and using ICT, graphs and diagrams to represent data including frequency diagrams for grouped discrete data	116
Interpreting and discussing results	To interpret diagrams and graphs, and draw simple conclusions based on the shape of graphs and simple statistics for a single distribution To compare two simple distributions using the range and one of the mode, median or mean To write a short report of a statistical enquiry and illustrate with appropriate diagrams, graphs and charts, using ICT as appropriate; justify the choice of what is presented	120–122
Probability	To use vocabulary and ideas of probability, drawing on experience To understand and use the probability scale from 0 to 1; find and justify probabilities based on equally likely outcomes in simple contexts To collect data from a simple experiment and record in a frequency table; estimate probabilities based on this data To compare experimental and theoretical probabilities in simple contexts	123–127

- Understand and use decimal notation and place value
- Present and interpret solutions in the context of the original problem …
- Suggest extensions to problems by asking 'What if …?'; begin to … significance of a counter-example

Place value

1 Read these whole numbers. Write the place value of each red digit.

a 5984	**b** 410	**c** 7162	**d** 19 891
e 21 987	**f** 736 019	**g** 6098	**h** 678 912
i 9 709 489	**j** 2 092 846	**k** 82 950	**l** 1 958 678

2 Read these decimal numbers. Write the place value of each red digit.

a 6·4	**b** 12·6	**c** 8·45	**d** 85·54
e 19·532	**f** 7·812	**g** 5·071	**h** 5·02
i 7·609	**j** 20·102	**k** 29·62	**l** 7·681

3 Write each of the following numbers in words.

a 198	**b** 4710	**c** 125 489	**d** 5·2
e 17·45	**f** 1 490 287	**g** 3·591	**h** 12·053

4 Write each of the following numbers in figures.

a two thousand, seven hundred and nine

b five point eight three

c one hundred and twelve

d zero point one five six

e two hundred and three thousand, one hundred and eleven

f six million, three hundred and forty thousand, one hundred and two

g eight point six zero seven

h six hundred and fifteen, point two three one

5 Write each of the following numbers in figures.

a 5 + 100 + 0·4 = ▢

b 7 + 20 + 0·5 + 600 = ▢

c 0·03 + 8 + 0·2 = ▢

d 900 + 6000 + 1 + 50 000 = ▢

e 6 + 0·008 + 0·7 + 0·01 = ▢

f 2 + 0·003 + 0·5 = ▢

g 100 + 6 000 000 + 2 + 70 000 = ▢

h 0·06 + 10 + 0·3 + 9 = ▢

3

a Using the digits on the cards, how many different numbers can you make? You must use all the digits each time. Remember to use whole numbers and decimal numbers.

b Now put all the numbers you have made in order, smallest to largest.

Multiplying and dividing by 10, 100 and 1000

1 Copy and complete.

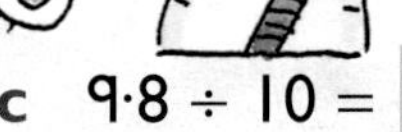
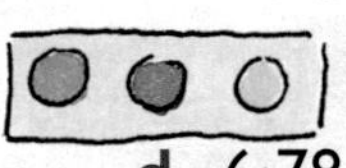
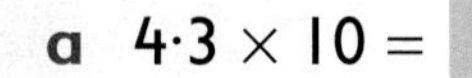

a 4·3 × 10 = ☐	**b** 16 ÷ 100 = ☐	**c** 9·8 ÷ 10 = ☐	**d** 6·78 × 100 = ☐
e 7·32 ÷ 1000 = ☐	**f** 9·2 × 1000 = ☐	**g** 6·25 ÷ 100 = ☐	**h** 4 ÷ 10 = ☐
i 8.1 × 100 = ☐	**j** 55·2 ÷ 10 = ☐	**k** 18·3 × 1000 = ☐	**l** 5·39 × 10 = ☐
m 34·89 × 100 = ☐	**n** 74·2 × 10 = ☐	**o** 16 ÷ 1000 = ☐	**p** 0·6 × 100 = ☐
q 682 ÷ 100 = ☐	**r** 95·7 ÷ 1000 = ☐	**s** 63·01 × 1000 = ☐	**t** 46·7 ÷ 10 = ☐

2 Write the missing numbers.

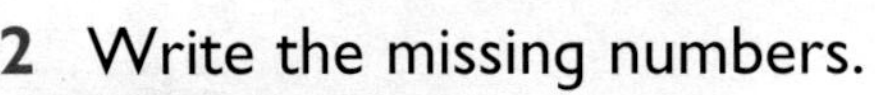
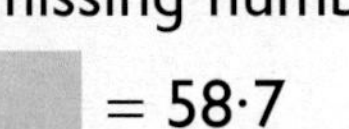

a 5·87 × ☐ = 58·7	**b** 81·8 ÷ ☐ = 0·818	**c** 202 ÷ ☐ = 0·202	**d** 0·12 × ☐ = 120
e 6·32 ÷ ☐ = 0·632	**f** 34·8 × ☐ = 348	**g** 9·32 × ☐ = 9320	**h** 0·4 × ☐ = 4
i 53·8 ÷ ☐ = 0·0538	**j** 92·2 ÷ ☐ = 9·22	**k** 0·25 × ☐ = 25	**l** 5·23 ÷ ☐ = 0·0523
m 8·59 × ☐ = 859	**n** 6·28 ÷ ☐ = 0·0628	**o** 5 ÷ ☐ = 0·5	**p** 34·7 × ☐ = 3470
q 1·8 ÷ ☐ = 0·018	**r** 32·58 × ☐ = 325·8	**s** 1·68 × ☐ = 1680	**t** 12 ÷ ☐ = 0·012

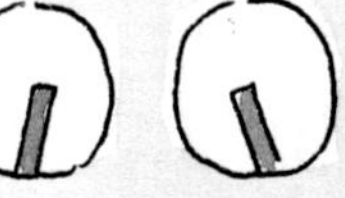
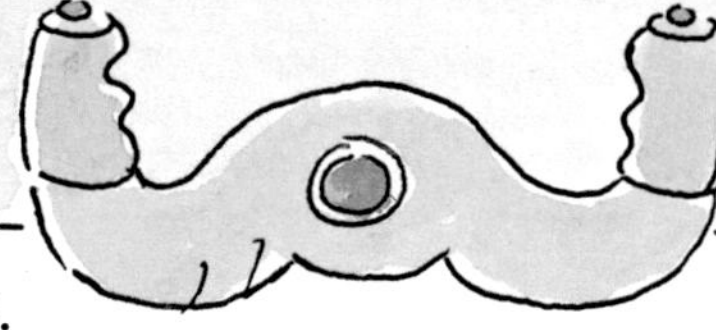
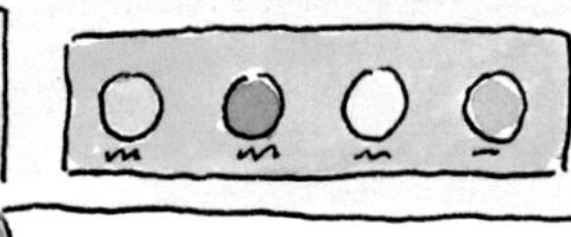
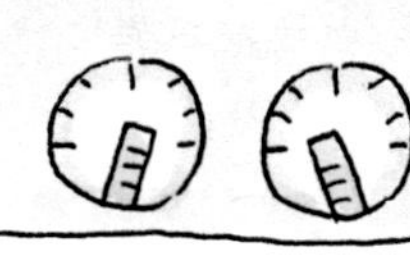

3 Solve these word problems.

a Madrid Primary is having a run-a-thon. Each child in Year 6 runs 100 laps around the school field. If each lap is 0·12 km, how far does each child run?

b Hot dog rolls come in packs of 10. The cost of 10 hot dogs is £7·20. How much does 1 hot dog cost?

c If 100 wooden beads weigh 2·84 kg, how much does each wooden bead weigh?

d A doctor spends on average 0·25 hours seeing each patient. Approximately how long does it take the doctor to see 10 patients?

Comparing and ordering decimals

1 Order each set of numbers, smallest to largest.

a 5·62, 5·26, 5·15, 5·49, 5·94, 5·38

b 7·42, 7·64, 7·41, 7·8, 7·48, 7·2

c 2·71, 2·51, 2·5, 2·02, 2·97, 2·17

d 6·233, 6·3, 6·323, 6·33, 6·303, 6·2

e 9·64, 9·602, 9·82, 9·06, 9·101, 9·72

f 32·7, 32·27, 32·74, 32·72, 32·47, 32·07

g 12·01, 12·1, 12·101, 12·12, 12·001, 12·21

h 26·755, 26·74, 26·703, 26·735, 26·7, 26·754

2 Order each set of measures, smallest to largest.

a 3·55 kg, 3·45 kg, 3·54 kg, 3·545 kg, 3·5 kg, 3·44 kg

b 1·6 l, 1·26 l, 1·06 l, 1·02 l, 1·12 l, 1·62 l

c 6·241 km, 6·41 km, 6·4 km, 6·421 km, 6·124 km, 6·14 km

d 3·3 cm, 3·8 cm, 3·383 cm, 3·83 cm, 3·38 cm, 3 cm

e 18·354 kg, 18·5 kg, 18·35 kg, 18·43 kg, 18·534 kg, 18 kg

f 14·4 l, 14·14 l, 14·41 l, 14·1 l, 14 l, 14·04 l

g 7·325 km, 7·3 km, 7·355 km, 7·53 km, 7·25 km, 7·5 km

h 25·2 m, 25·25 m, 25·643 m, 25·253 m, 25·52 m, 25·235 m

Puzzle time

a Using each of these digit cards and a decimal point card, investigate how many different decimal numbers you can make. You must use each card every time. Remember to use decimals with 1, 2 and 3 decimal places.

b Now put all the decimals you have made in order, smallest to largest.

Ordering and comparing measures

1 Copy and complete.

a	0·437 km = ☐ m	**b**	8·2 l = ☐ ml	**c**	6·42 m = ☐ cm	**d**	4370 cm = ☐ km
e	8000 ml = ☐ l	**f**	832 g = ☐ kg	**g**	0·52 kg = ☐ g	**h**	1·78 l = ☐ cl
i	10 ml = ☐ cl	**j**	4 km = ☐ cm	**k**	7500 kg = ☐ tonnes	**l**	6·5 gallons = ☐ pints
m	40 oz = ☐ lb	**n**	260 cl = ☐ l	**o**	26 cm = ☐ m	**p**	16 800 m = ☐ km
q	3·5 mm = ☐ cm	**r**	1·9 tonnes = ☐ kg	**s**	8·2 cm = ☐ mm	**t**	120 mm = ☐ m

2 Order each set of metric measures, smallest to largest.

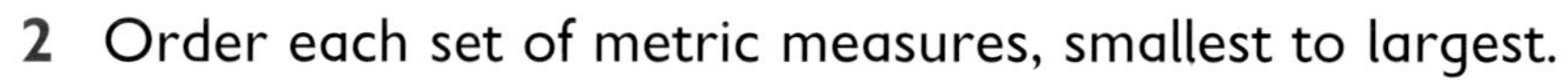

a 0·0013 km, 1100 mm, 120 cm, 1·6 m, 1800 mm, 150 cm

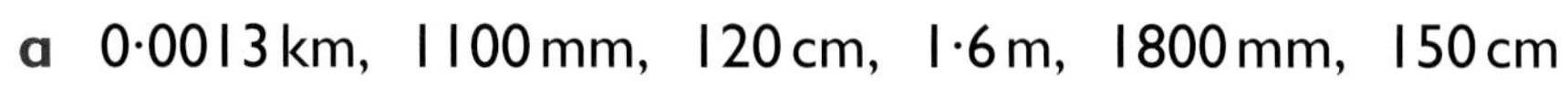

b 6·9 kg, 0·0096 tonnes, 5400 g, 0·9 kg, 0·0023 tonnes, 8300 g

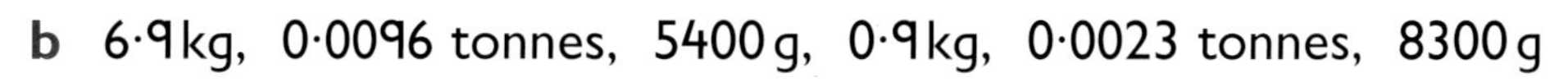

c 540 cl, 1·2 l, 2·7 l, 180 cl, 3600 ml, 980 ml

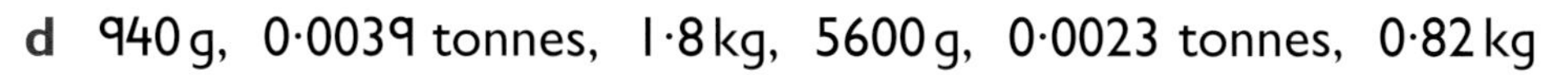

d 940 g, 0·0039 tonnes, 1·8 kg, 5600 g, 0·0023 tonnes, 0·82 kg

e 280 cl, 1·5 l, 4200 ml, 0·87 l, 63 cl, 70 ml

f 630 mm, 0·98 m, 1300 mm, 0·0011 km, 87 cm, 140 cm

3 Order each set of metric and imperial measures, smallest to largest.

a 5100 mm, 0·003 miles, 0·0015 miles, 0·0041 km, 340 cm, 6·3 m

b 7·1 l, 240 cl, 1·5 pints, 0·75 gallons, 11 pints, 4800 ml

c 2400 g, 100 oz, 2·5 lb, 0·98 kg, 0·0049 tonnes, 25 oz

d 0·25 gallons, 1·5 gallons, 170 cl, 5200 ml, 1 pint, 0·78 l

e 1·8 m, 280 cm, 0·0025 miles, 4300 mm, 0·0015 miles, 0·0032 km

f 2·4 kg, 3·5 lb, 4300 g, 90 oz, 7·6 lb, 0·0029 tonnes

Puzzle time

If two weights are interchanged at each move, what are the fewest moves needed to place these weights in order from lightest to heaviest?

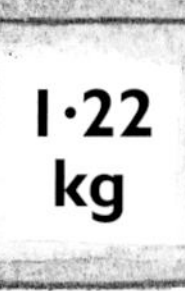

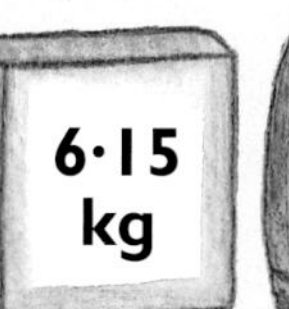

Rounding numbers

Example
4578 = 4580 to the nearest 10
4600 to the nearest 100
5000 to the nearest 1000

1 Round each of these numbers to the nearest 10, 100 and 1000.

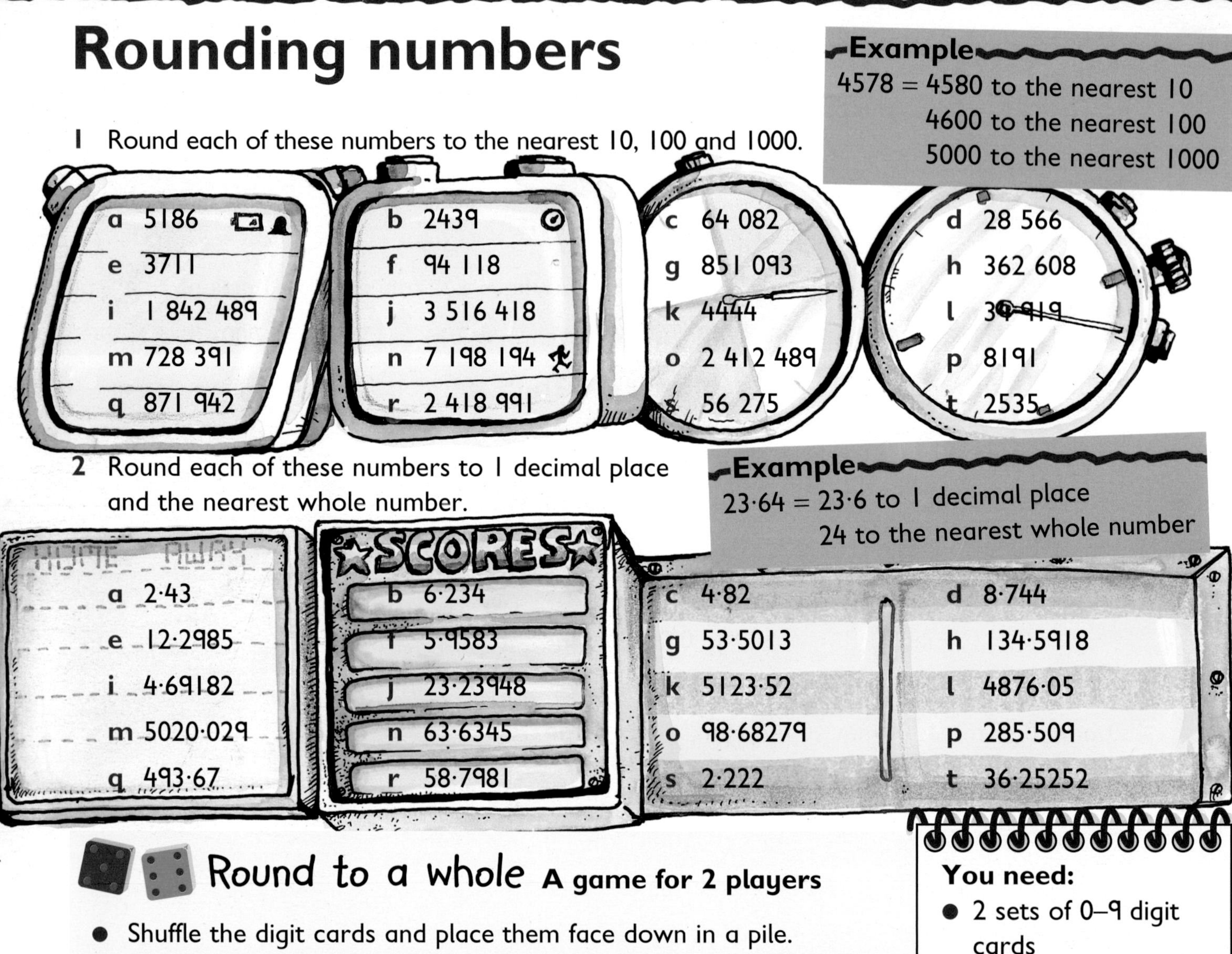

a 5186
b 2439
c 64 082
d 28 566
e 3711
f 94 118
g 851 093
h 362 608
i 1 842 489
j 3 516 418
k 4444
l 39 919
m 728 391
n 7 198 194
o 2 412 489
p 8191
q 871 942
r 2 418 991
s 56 275
t 2535

2 Round each of these numbers to 1 decimal place and the nearest whole number.

Example
23·64 = 23·6 to 1 decimal place
24 to the nearest whole number

a 2·43
b 6·234
c 4·82
d 8·744
e 12·2985
f 5·9583
g 53·5013
h 134·5918
i 4·69182
j 23·23948
k 5123·52
l 4876·05
m 5020·029
n 63·6345
o 98·68279
p 285·509
q 493·67
r 58·7981
s 2·222
t 36·25252

Round to a whole **A game for 2 players**

You need:
- 2 sets of 0–9 digit cards
- 2 decimal point cards

- Shuffle the digit cards and place them face down in a pile.
- Take turns to choose 3 digit cards and, using the decimal point card, make a number with 2 decimal places, for example, 5·72.
- Round your decimal to the nearest whole number, i.e. 6. That is your score for that round.
- After 10 rounds add up your scores. The winner is the player with the larger total.

Variation:
- Take turns to choose 4 digit cards to make a number with 3 decimal places, for example, 6·498.

Round to a tenth **A game for 2 players**

- Shuffle the digit cards and place them face down in a pile.
- Take turns to choose 3 digit cards and, using the decimal point card, make a number with 2 decimal places, for example, 5·72.
- Round your decimal to the nearest tenth, i.e. 5·7. This is your score for that round.
- After 10 rounds add up your scores. The winner is the player with the larger total.

Variation:
- Take turns to choose 4 digit cards to make a number with 3 decimal places, for example, 6·498.

Positive and negative numbers

1 Place each set of numbers in order, smallest to largest.

a $^{+}5$, $^{-}9$, $^{+}2$, $^{-}6$, $^{+}9$, $^{-}5$

b $^{-}12$, $^{+}8$, $^{-}14$, $^{-}7$, $^{+}2$, $^{-}10$

c $^{+}9$, $^{-}7$, $^{+}1$, $^{-}11$, $^{+}4$, $^{-}9$

d $^{-}15$, $^{+}23$, $^{-}26$, $^{+}25$, $^{-}19$, $^{+}5$

e $^{-}23$, $^{-}19$, $^{+}35$, $^{-}39$, $^{+}15$, $^{-}29$

f $^{+}54$, $^{+}45$, $^{-}19$, $^{+}55$, $^{-}29$, $^{+}34$

g $^{+}64$, $^{-}46$, $^{+}46$, $^{-}63$, $^{+}36$, $^{-}45$

h $^{-}31$, $^{+}32$, $^{-}33$, $^{-}34$, $^{+}35$, $^{-}36$

i $^{+}37$, $^{-}18$, $^{+}29$, $^{-}16$, $^{+}24$, $^{-}21$

j $^{-}45$, $^{-}57$, $^{+}33$, $^{-}52$, $^{+}35$, $^{-}43$

2 Write < or > signs between each set of numbers.

a $^{-}15$ ☐ $^{+}15$ ☐ $^{-}5$

b $^{+}33$ ☐ $^{-}17$ ☐ $^{+}24$

c $^{-}9$ ☐ $^{-}12$ ☐ $^{+}11$

d $^{-}18$ ☐ $^{+}23$ ☐ $^{-}10$ ☐ $^{+}12$

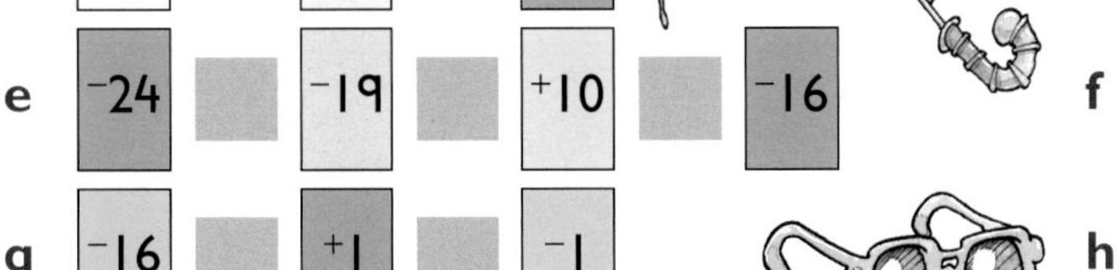

e $^{-}24$ ☐ $^{-}19$ ☐ $^{+}10$ ☐ $^{-}16$

f $^{+}29$ ☐ $^{+}16$ ☐ $^{-}12$

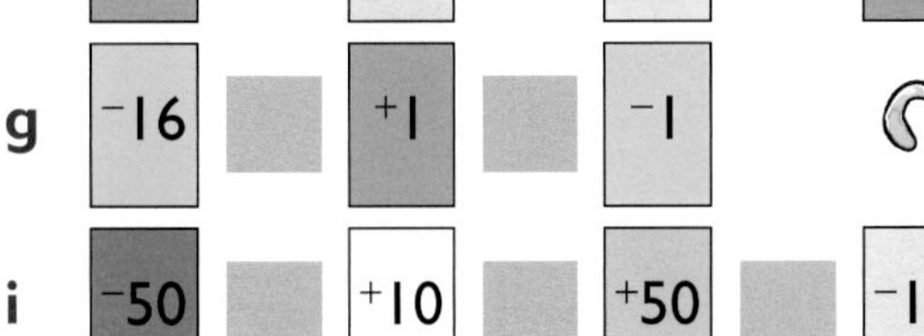

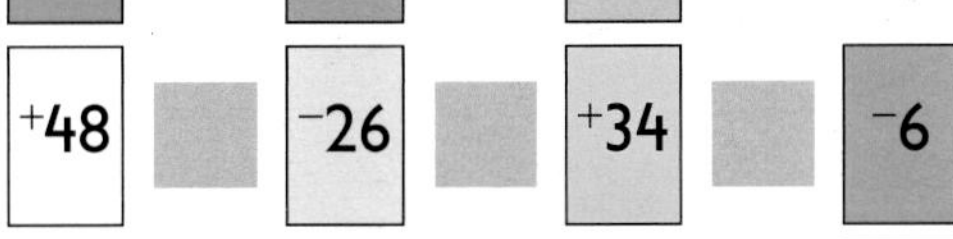

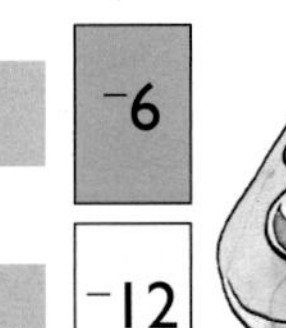

g $^{-}16$ ☐ $^{+}1$ ☐ $^{-}1$

h $^{+}48$ ☐ $^{-}26$ ☐ $^{+}34$ ☐ $^{-}6$

i $^{-}50$ ☐ $^{+}10$ ☐ $^{+}50$ ☐ $^{-}10$

j $^{+}35$ ☐ $^{+}30$ ☐ $^{-}22$ ☐ $^{-}12$

Order the numbers **A game for 2 players**

You need:
- a set of number cards from Resource Copymasters 3 and 4 (per pair)

- Shuffle the cards and place them face down in a pile.
- Player 1 turns over six cards and places them face up in the order in which they were chosen from the pack, for example:

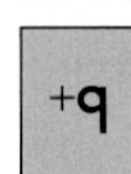
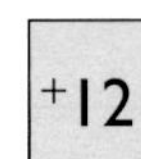

$^{+}9$ $^{+}12$ $^{-}17$ $^{-}6$ $^{-}10$ $^{+}18$

- Player 1 now swaps the position of two cards at a time until all the cards are in order, smallest to largest, for example:

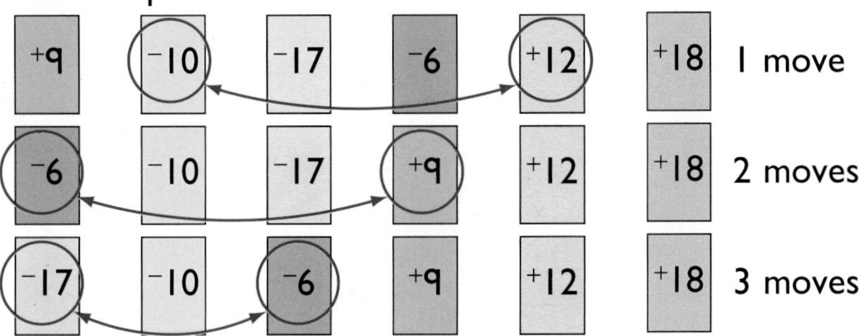

$^{+}9$ $^{-}10$ $^{-}17$ $^{-}6$ $^{+}12$ $^{+}18$ 1 move

$^{-}6$ $^{-}10$ $^{-}17$ $^{+}9$ $^{+}12$ $^{+}18$ 2 moves

$^{-}17$ $^{-}10$ $^{-}6$ $^{+}9$ $^{+}12$ $^{+}18$ 3 moves

- Player 2 counts the number of moves Player 1 took to place all the cards in order, in this case, 3 moves.
- Player 1 scores 3 points for that round.
- Swap roles.
- After 5 rounds each player finds the total of all their moves.
- The overall winner is the player with fewer moves.

- Add and subtract positive and negative integers in context
- Present and interpret solutions …; explain and justify methods and conclusions …
- Suggest extensions to problems by asking 'What if …?'; begin to … significance of a counter-example

Adding and subtracting positive and negative integers

Help the spy work out the secret message. He already knows that the secret message has 12 words. Answer each set of calculations. Then using the Spy Master's Secret Code, work out which letter matches each answer to decipher the message.

SPY MASTER'S SECRET CODE

$^{+}9$	$^{-}1$	$^{+}4$	$^{+}12$	$^{-}3$	$^{-}8$	$^{+}1$	$^{-}2$	$^{+}8$	$^{-}5$	$^{+}6$	$^{-}9$	$^{+}13$
A	B	C	D	E	F	G	H	I	J	K	L	M
$^{-}7$	$^{-}4$	$^{-}11$	$^{+}10$	$^{+}5$	$^{+}3$	$^{-}10$	$^{-}12$	$^{-}6$	$^{+}11$	$^{-}13$	$^{+}2$	$^{+}7$
N	O	P	Q	R	S	T	U	V	W	X	Y	Z

1
$^{-}5 + ^{-}5 =$
$^{+}3 + ^{-}5 =$
$^{-}5 + ^{+}2 =$

2
$^{+}5 - ^{-}7 =$
$^{-}1 + ^{-}3 =$
$^{+}3 - ^{-}1 =$
$^{-}7 + ^{-}5 =$
$^{+}8 - ^{-}5 =$
$0 + ^{-}3 =$
$^{-}4 + ^{-}3 =$
$^{-}5 - ^{+}5 =$
$^{+}8 + ^{-}5 =$

3
$^{-}6 + ^{+}15 =$
$^{-}7 + ^{+}12 =$
$^{+}6 + ^{-}9 =$

4
$^{-}6 + ^{-}6 =$
$^{+}8 + ^{-}15 =$
$^{+}8 - ^{-}4 =$
$^{-}7 + ^{+}4 =$
$^{-}9 + ^{+}14 =$

5
$^{-}2 + ^{-}8 =$
$^{+}8 + ^{-}10 =$
$^{-}10 + ^{+}7 =$

6
$^{-}6 + ^{+}5 =$
$^{+}12 + ^{-}15 =$
$^{-}1 + ^{-}6 =$
$^{+}5 + ^{+}3 + ^{-}4 =$
$^{+}12 + ^{-}14 =$

7
$^{+}3 + ^{-}4 =$
$^{+}6 + ^{-}8 + ^{-}1 =$
$^{+}14 + ^{-}7 - ^{+}4 =$
$^{-}6 - ^{-}5 + ^{+}9 =$
$^{-}3 - ^{-}17 - ^{+}2 =$
$^{+}8 - ^{+}14 + ^{+}3 =$

8
$^{+}3 - ^{+}13 =$
$^{-}13 - ^{-}15 - ^{+}4 =$
$^{+}17 - ^{+}20 =$

9
$^{+}11 - ^{+}18 + ^{+}6 =$
$^{+}14 - ^{-}9 + ^{-}18 =$
$^{-}9 + ^{+}17 =$
$^{-}5 - ^{-}19 + ^{-}2 =$
$^{-}16 + ^{+}8 - ^{-}9 =$
$^{+}13 + ^{-}7 - ^{+}9 =$

10
$^{-}21 + ^{+}17 =$
$^{+}9 + ^{-}22 - ^{-}7 =$
$^{+}14 + ^{-}27 + ^{+}10 =$
$^{-}23 - ^{-}28 =$

11
$^{+}14 - ^{-}13 - ^{+}22 =$
$^{-}4 - ^{-}29 + ^{-}17 =$
$^{-}28 + ^{+}12 - ^{-}10 =$
$^{+}16 + ^{-}27 + ^{+}8 =$
$^{-}21 + ^{+}26 =$

12
$^{+}4 + ^{+}3 + ^{-}4 =$
$^{+}33 - ^{+}35 =$
$^{+}24 - ^{+}15 + ^{-}13 =$
$^{+}36 - ^{+}27 =$
$^{-}16 + ^{+}33 - ^{+}26 =$
$^{+}36 - ^{+}17 + ^{-}21 =$
$^{-}16 + ^{-}2 + ^{+}27 =$
$^{-}12 - ^{+}5 - ^{-}11 =$
$^{-}7 + ^{+}29 - ^{+}25 =$
$^{+}11 - ^{+}6 + ^{-}12 =$

Multiples, factors and tests of divisibility

You need:
- a 1–10 dice (or a 0–9 dice with a 1 written in front of the 0 to represent 10)

1 Choose a number from the grid, for example, 376 and write it down. Now roll the dice, for example, 7. Work out if the number from the grid is divisible by the number rolled on the dice. If the number is divisible draw a tick (✓) beside it, if not, draw a cross (✗). Your teacher will tell you how many numbers to choose.

Example
376 is divisible by 7 ✗

523	5146	142	287	890	4556
618	34	7351	43	3938	62
71	9248	66	13 495	89	72 455
759	1677	376	864	8352	497
83	24 128	6269	7679	924	21
60 716	32	804	2580	46 907	75

2 Write all the pairs of factors of each number.

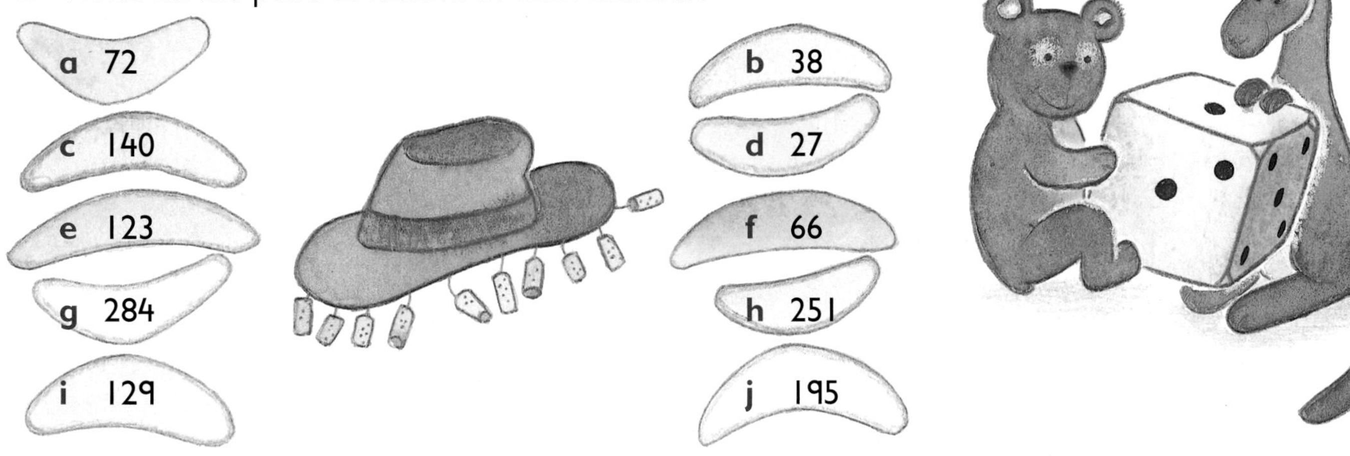

3 Choose 10 numbers from the grid above and write all the pairs of factors.

Common factors

1 Write all the common factors for each pair of numbers.

a	6 and 8	**b**	14 and 28
c	42 and 54	**d**	36 and 63
e	104 and 144	**f**	147 and 138
g	184 and 272	**h**	165 and 240
i	256 and 288	**j**	222 and 246
k	250 and 340	**l**	369 and 216

2 Draw a ring around the highest common factor for each pair of numbers in question 1.

3 Choose two numbers from the octagon and write down all the common factors for each pair. Draw a ring around the highest common factor. Your teacher will tell you how many pairs of numbers to choose.

4 Look at the octagon in question 3. How many sets of 3 numbers have common factors?

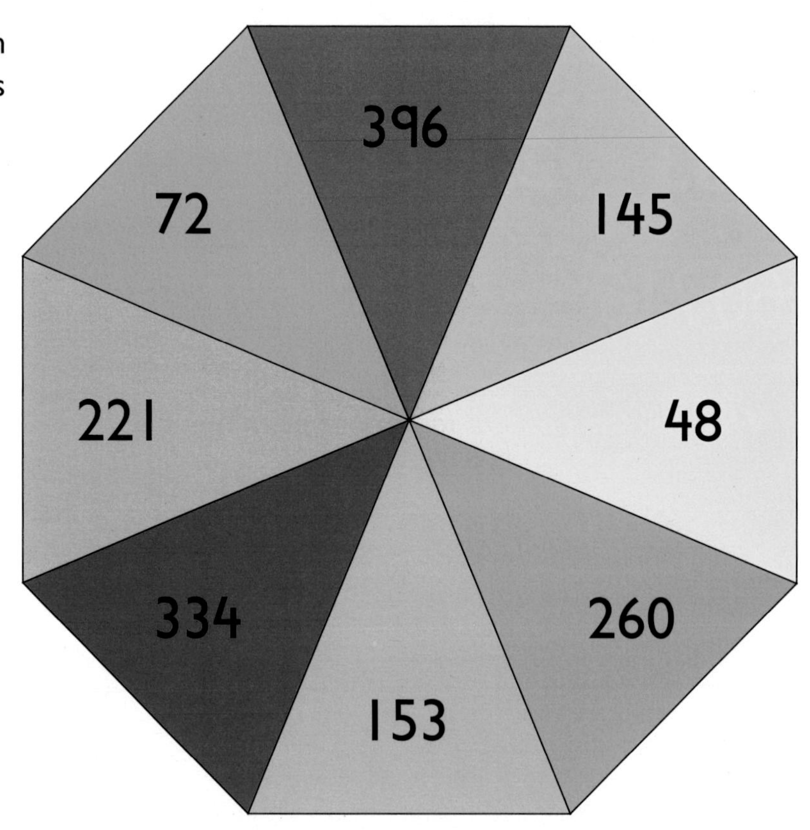

Multiples

1 Find all the common multiples up to 100 for each pair of numbers.

a 4 and 6

b 3 and 9

c 2 and 5

d 3 and 7

e 7 and 8

f 3 and 12

g 6 and 9

h 6 and 18

i 3 and 10

j 9 and 24

k 4 and 11

l 10 and 25

2 Draw a ring around the lowest common multiple for each pair of numbers in question 1.

3 Copy and complete. Write a number in each circle so that the number in each square is the product of the two numbers on either side.

a

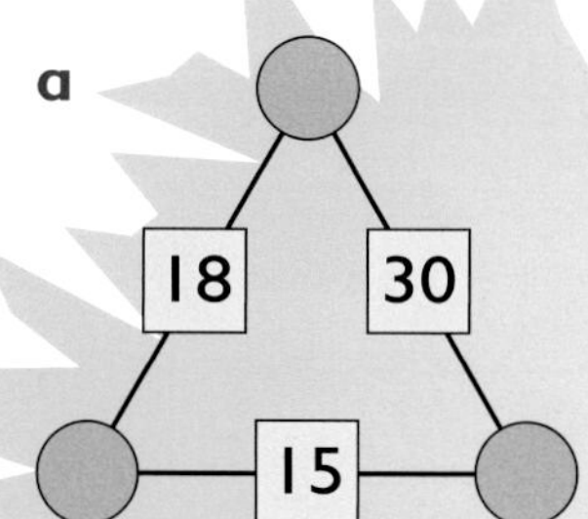

b

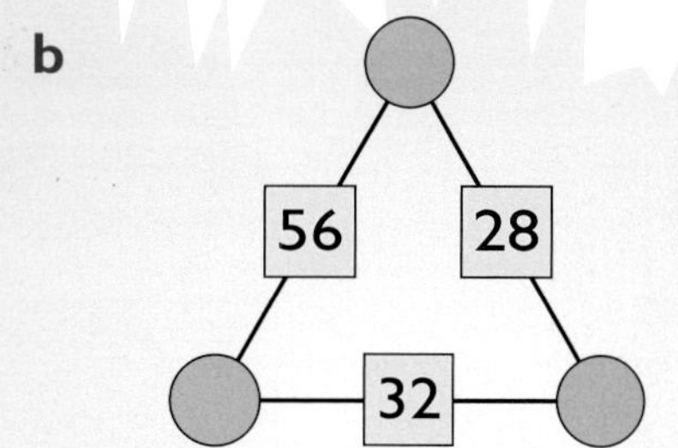

c

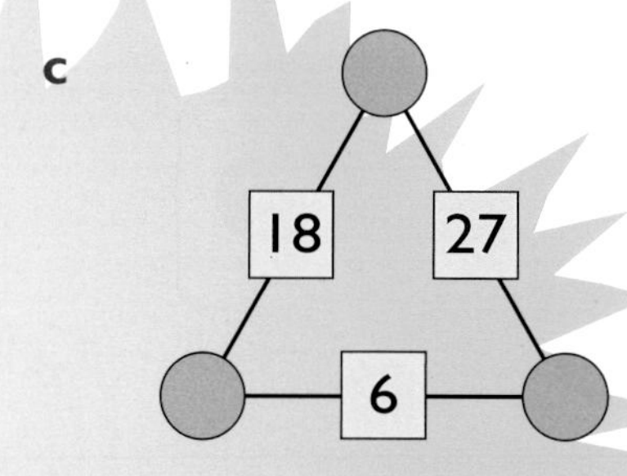

d

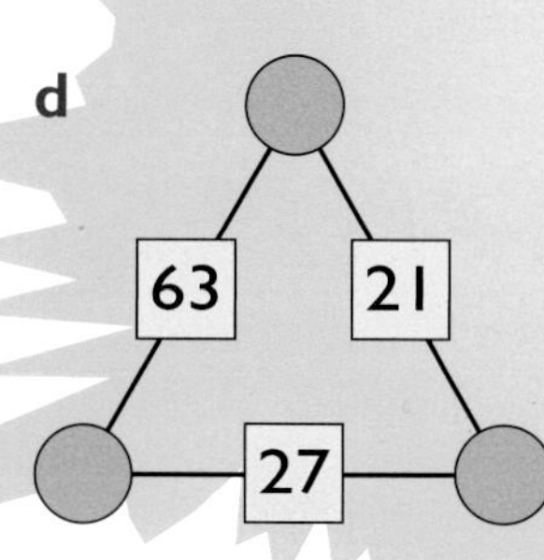

e

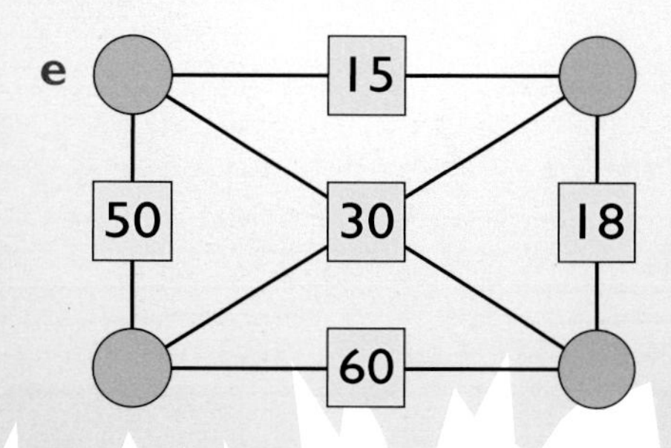

f

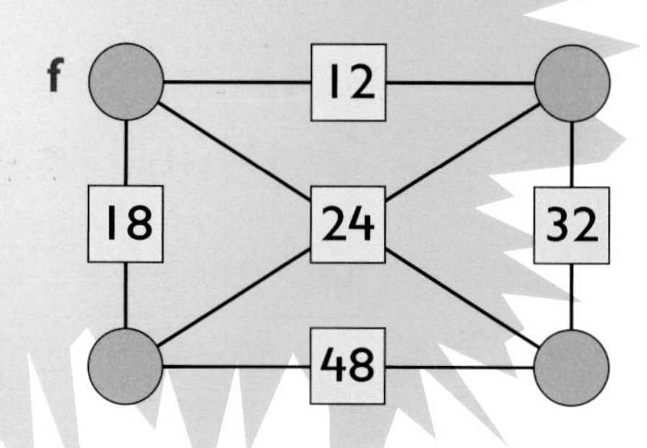

4 Make a question like those above for a friend to solve.

Prime numbers

1 Write the prime numbers that come before and after each of these numbers.

a 18	b 46	c 62	d 80
e 30	f 72	g 85	h 34
i 39	j 27	k 55	l 92
m 49	n 68	o 20	p 60

Puzzle time

2 Solve each of these puzzles.

a Find all the prime numbers to 100 which, when their digits are reversed, are also prime.

b Which prime numbers to 100 can be written as the sum of two square numbers?

c 3 and 5 are prime numbers. They are also consecutive odd numbers. Find how many pairs of consecutive odd numbers to 100 are both prime.

Gather the primes

A game for 2 players

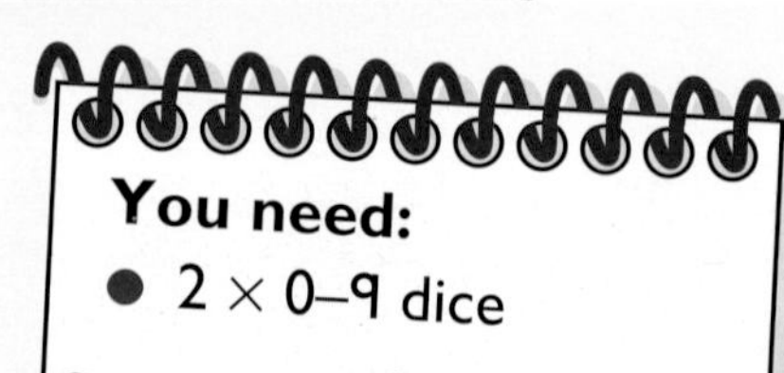

You need:
- 2 × 0–9 dice

- Take turns to roll the dice, for example, 4 and 3, and make a two-digit number, i.e. 43 or 34.
- If the number is a prime number write it down.
- The winner is the first player to write down 10 prime numbers.

- Recognise the first few triangular numbers and squares of numbers to at least 12×12
- Present and interpret solutions …; explain and justify methods …
- Suggest extensions … by asking 'What if …?'; begin to generalise … of a counter-example

Lesson 12

Triangular numbers and squares

1 Copy and complete.

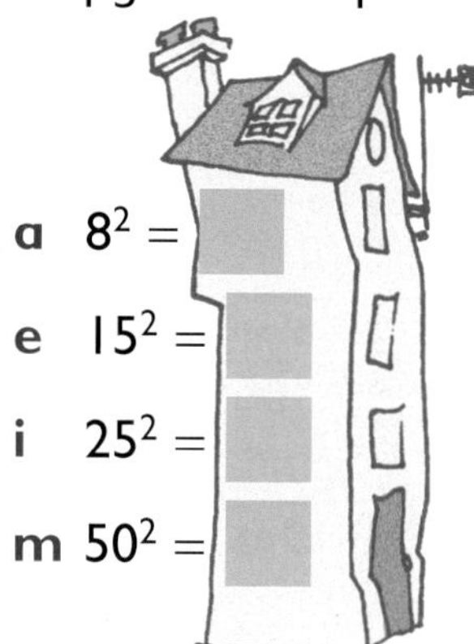
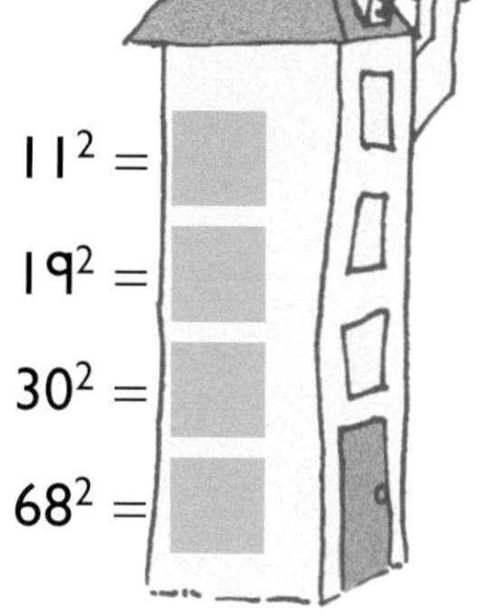
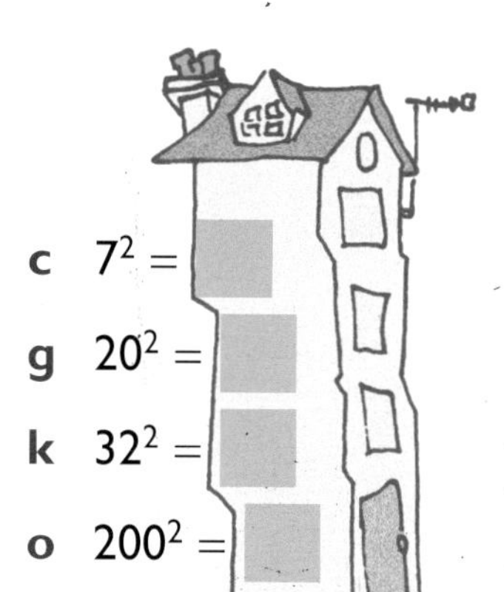
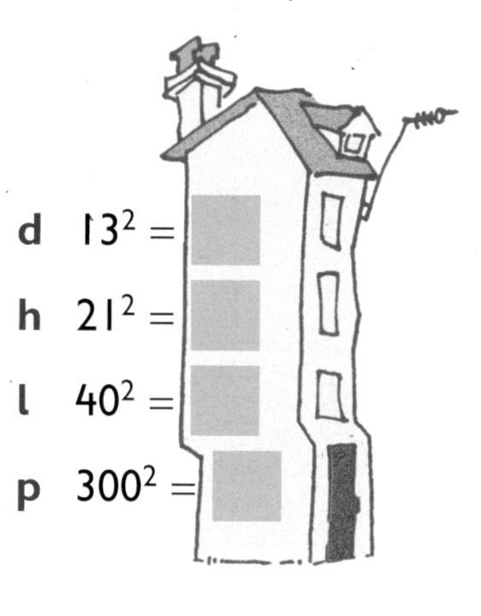

a $8^2 =$ | b $11^2 =$ | c $7^2 =$ | d $13^2 =$

e $15^2 =$ | f $19^2 =$ | g $20^2 =$ | h $21^2 =$

i $25^2 =$ | j $30^2 =$ | k $32^2 =$ | l $40^2 =$

m $50^2 =$ | n $68^2 =$ | o $200^2 =$ | p $300^2 =$

2 Copy and complete.

a $8^2 + 7^2 =$ | b $12^2 + 5^2 =$ | c $6^2 - 3^2 =$ | d $9^2 - 4^2 =$

e $10^2 - 5^2 =$ | f $14^2 + 2^2 =$ | g $16^2 - 8^2 =$ | h $18^2 + 10^2 =$

i $20^2 - 14^2 =$ | j $22^2 + 1^2 =$ | k $11^2 + 24^2 =$ | l $21^2 - 12^2 =$

m $30^2 + 17^2 =$ | n $40^2 - 20^2 =$ | o $15^2 + 17^2 - 12^2 =$ | p $22^2 - 13^2 + 17^2 =$

q $5^2 \times 4^2 =$ | r $9^2 \times 12^2 =$ | s $6^2 \div 3^2 =$ | t $27^2 \div 9^2 =$

Puzzle time

Solve each of these puzzles.

a Without using a calculator find a number that when multiplied by itself gives 2304.

b Which triangular numbers are equal to the sum of two other triangular numbers?

c Can every square number up to 12×12 be expressed as the sum of two prime numbers?

d Look at the pattern formed by the last digit of square numbers. Do any numbers not appear as the last digit? Could 517 be a square number? Or 326?

e Choose two consecutive triangular numbers and add them together. What do you notice about the total?

Squares and roots

1 Work out the square roots without using a calculator.

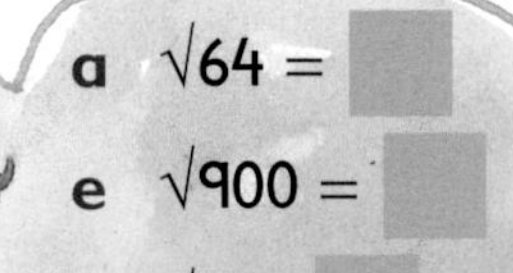

a $\sqrt{64}$ = ☐	**b** $\sqrt{81}$ = ☐	**c** $\sqrt{144}$ = ☐	**d** $\sqrt{1}$ = ☐
e $\sqrt{900}$ = ☐	**f** $\sqrt{484}$ = ☐	**g** $\sqrt{324}$ = ☐	**h** $\sqrt{625}$ = ☐
i $\sqrt{9}$ = ☐	**j** $\sqrt{1600}$ = ☐	**k** $\sqrt{100}$ = ☐	**l** $\sqrt{121}$ = ☐
m $\sqrt{25}$ = ☐	**n** $\sqrt{169}$ = ☐	**o** $\sqrt{10\,000}$ = ☐	**p** $\sqrt{90\,000}$ = ☐

2 Estimate these square roots. Then use a calculator to check, rounding the answer to two decimal places.

Example
$\sqrt{175} \approx 14 = 13{\cdot}23$

You need:
- a calculator

a $\sqrt{162} \approx$ ☐ = ☐	**b** $\sqrt{184} \approx$ ☐ = ☐	**c** $\sqrt{156} \approx$ ☐ = ☐	**d** $\sqrt{236} \approx$ ☐ = ☐
e $\sqrt{275} \approx$ ☐ = ☐	**f** $\sqrt{294} \approx$ ☐ = ☐	**g** $\sqrt{378} \approx$ ☐ = ☐	**h** $\sqrt{469} \approx$ ☐ = ☐
i $\sqrt{481} \approx$ ☐ = ☐	**j** $\sqrt{502} \approx$ ☐ = ☐	**k** $\sqrt{687} \approx$ ☐ = ☐	**l** $\sqrt{693} \approx$ ☐ = ☐
m $\sqrt{730} \approx$ ☐ = ☐	**n** $\sqrt{884} \approx$ ☐ = ☐	**o** $\sqrt{921} \approx$ ☐ = ☐	**p** $\sqrt{2400} \approx$ ☐ = ☐
q $\sqrt{3700} \approx$ ☐ = ☐	**r** $\sqrt{5000} \approx$ ☐ = ☐	**s** $\sqrt{12000} \approx$ ☐ = ☐	**t** $\sqrt{1300} \approx$ ☐ = ☐

Estimate the square

A game for 2 players

You need:
- 2 × 1–6 dice
- a calculator

- Take turns to roll the dice, for example, 5 and 4 and make a two-digit number, i.e. 45 or 54.
- Each player estimates the square of the number and writes it down.
- Now check the answer with a calculator.
- The player with the nearer estimate draws a ring around their estimate.
- The overall winner is the first player to have drawn a ring around 5 estimates.

Variation:
Try using two 0–9 dice.

Estimate the square root

A game for 2 players

You need:
- 3 × 1–6 dice
- a calculator

- Take turns to roll the dice, for example, 6, 3 and 4 and make a three-digit number, i.e. 346.
- Each player estimates the square root of the number and writes it down.
- Now check the answer with a calculator.
- The player with the nearer estimate draws a ring around their estimate.
- The overall winner is the first player to have drawn a ring around 5 estimates.

Variation:
Try using three 0–9 dice.

Fractions

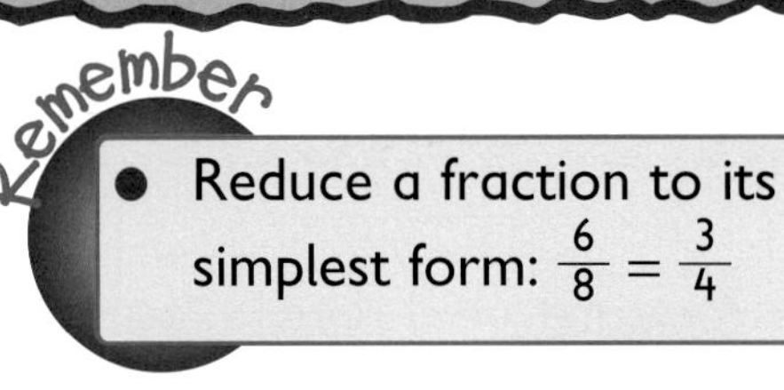

- Reduce a fraction to its simplest form: $\frac{6}{8} = \frac{3}{4}$

1 What fraction of the large shape is the small shape?

a 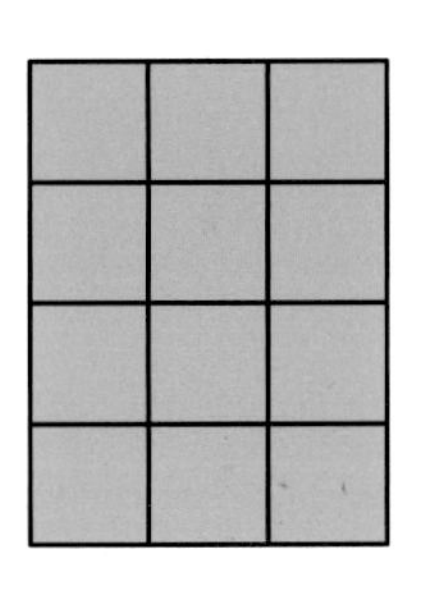

b 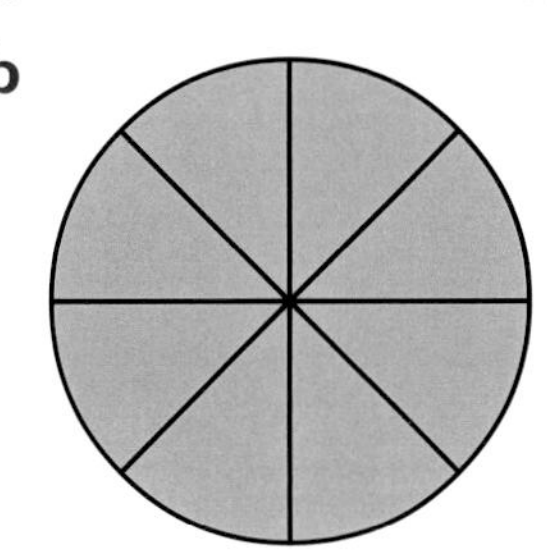 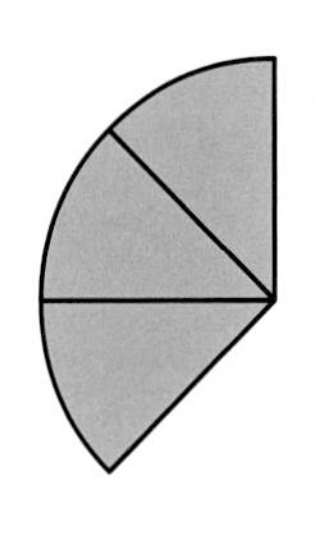

c 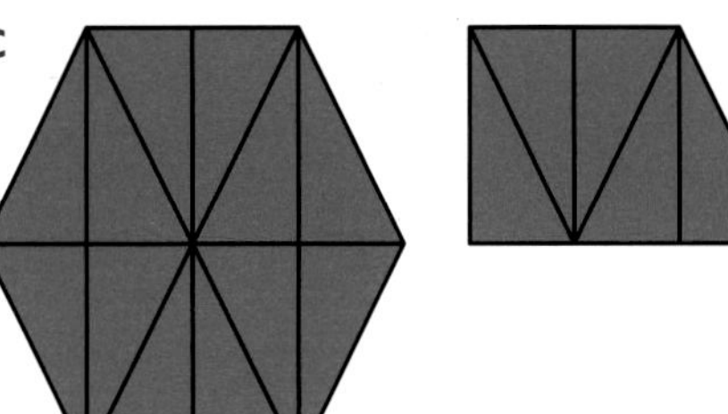

2 What fraction of a complete turn does the minute hand turn between:

a 2:30 a.m. and 2:45 a.m.?

b 9:10 a.m. and 9:40 a.m.?

c 2:25 p.m. and 2:45 p.m.?

d 8:10 p.m. and 8:55 p.m.?

e 6:55 a.m. and 7:05 a.m.?

3 What fraction of…

a I metre is 60 centimetres?

b I day is 4 hours?

c I kilogram is 37 grams?

d I kilometre is 270 metres?

e I week is the weekend?

f I litre is 8 centilitres?

g I hour is 35 minutes?

h I gallon is 2 pints?

i I pound is 2 ounces?

j I year is 4 months?

k I litre is 300 millilitres?

l I kilogram is 630 grams?

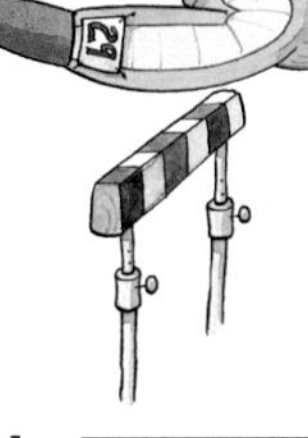

4 Estimate the fraction of each shape that is shaded.

a 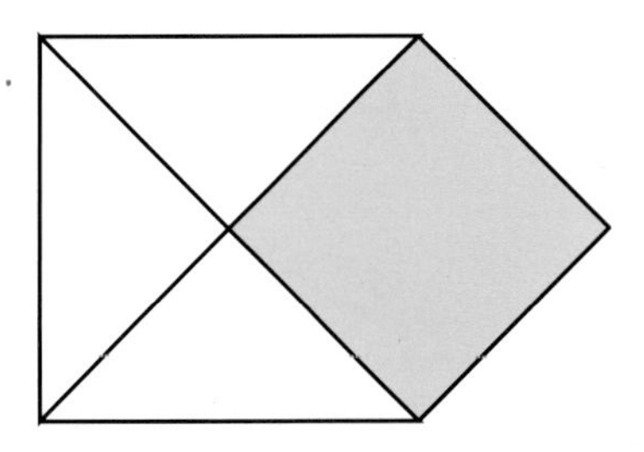

b 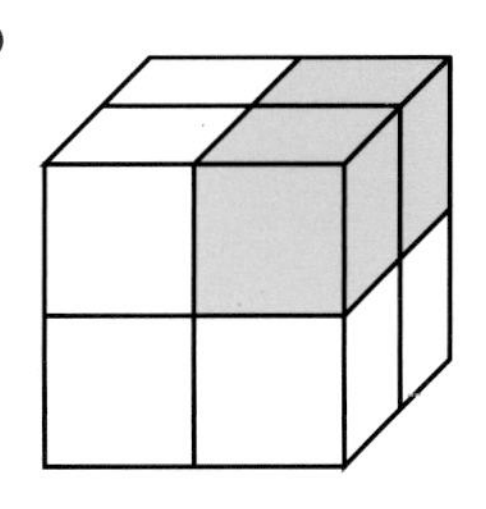

c 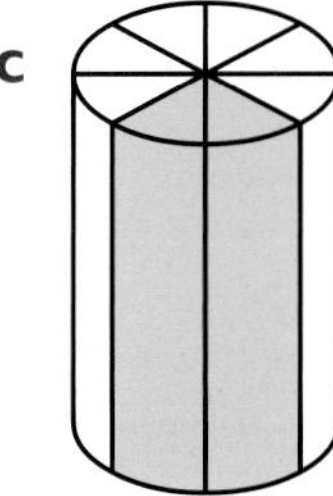

d 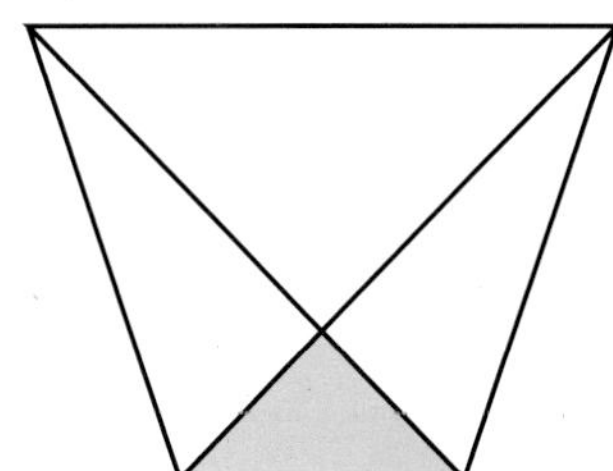

5 Investigate how many different ways you can divide this grid into quarters using straight lines. Here is one way:

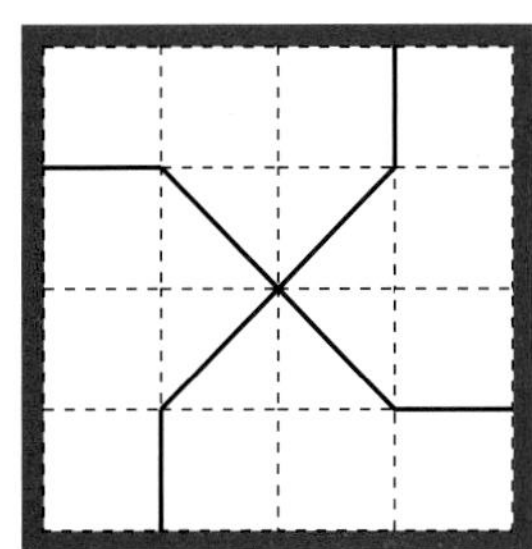

Equivalent fractions

1 Complete the pairs of equivalent fractions.

a $\frac{3}{5} = \frac{\square}{20}$	b $\frac{4}{14} = \frac{\square}{28}$	c $\frac{\square}{9} = \frac{6}{27}$	d $\frac{\square}{6} = \frac{30}{36}$
e $\frac{8}{\square} = \frac{32}{80}$	f $\frac{3}{\square} = \frac{21}{49}$	g $\frac{6}{8} = \frac{36}{\square}$	h $\frac{1}{3} = \frac{7}{\square}$
i $\frac{\square}{4} = \frac{48}{64}$	j $\frac{2}{16} = \frac{8}{\square}$	k $\frac{9}{\square} = \frac{63}{70}$	l $\frac{1}{45} = \frac{\square}{90}$
m $\frac{3}{13} = \frac{15}{\square}$	n $\frac{\square}{12} = \frac{40}{96}$	o $\frac{4}{\square} = \frac{24}{600}$	p $\frac{\square}{100} = \frac{56}{700}$
q $\frac{3}{17} = \frac{\square}{34}$	r $\frac{6}{\square} = \frac{7}{28}$	s $\frac{\square}{21} = \frac{24}{56}$	t $\frac{12}{32} = \frac{18}{\square}$

2 Reduce these fractions to their simplest form.

a $\frac{12}{14}$	b $\frac{6}{36}$	c $\frac{7}{28}$	d $\frac{30}{45}$
e $\frac{16}{36}$	f $\frac{12}{20}$	g $\frac{72}{81}$	h $\frac{8}{20}$
i $\frac{3}{21}$	j $\frac{18}{24}$	k $\frac{21}{49}$	l $\frac{24}{64}$
m $\frac{15}{24}$	n $\frac{20}{28}$	o $\frac{28}{63}$	p $\frac{24}{50}$
q $\frac{75}{100}$	r $\frac{24}{30}$	s $\frac{39}{78}$	t $\frac{36}{81}$

Puzzle time

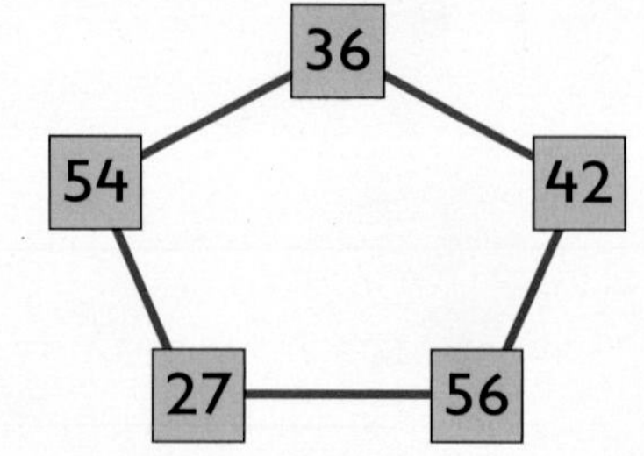

a Using two adjoining numbers, one as the numerator, the other as a denominator, make fractions equivalent to $\frac{1}{2}$, $\frac{2}{3}$ and $\frac{3}{4}$.

b Look at this fraction: $\frac{26}{65}$

If you cross out the two digits that are the same you create an equivalent fraction. $\frac{26}{65} = \frac{2}{5}$. Find other pairs of fractions where this is true.

Converting decimals and fractions

1 Convert these decimals to fractions in their simplest form.

a 1·5 = 	b 4·25 = 	c 0·4 = 	d 4·17 =

e 0·3 = 	f 3·7 = 	g 7·75 = 	h 5·35 =

i 0·24 = 	j 0·6 = 	k 0·28 = 	l 3·64 =

m 4·85 = 	n 17·32 = 	o 3·125 = 	p 5·04 =

q 0·48 = 	r 0·25 = 	s 3·65 = 	t 0·36 =

2 Convert these fractions to decimals.

a $\frac{5}{20} =$ 	b $\frac{7}{21} =$ 	c $3\frac{2}{5} =$ 	d $\frac{12}{18} =$

e $\frac{18}{24} =$ 	f $\frac{3}{8} =$ 	g $\frac{12}{16} =$ 	h $5\frac{9}{15} =$

i $\frac{30}{40} =$ 	j $\frac{24}{30} =$ 	k $\frac{16}{32} =$ 	l $\frac{42}{60} =$

m $\frac{9}{27} =$ 	n $\frac{7}{35} =$ 	o $\frac{90}{300} =$ 	p $\frac{5}{8} =$

q $2\frac{4}{5} =$ 	r $\frac{36}{40} =$ 	s $\frac{3}{5} =$ 	t $\frac{10}{100} =$

Comparing fractions

1 Use the $<$ and $>$ signs to make each statement correct.

a	$\frac{5}{6}$ ☐ $\frac{2}{3}$	**b**	$\frac{1}{2}$ ☐ $\frac{7}{10}$	**c**	$\frac{3}{4}$ ☐ $\frac{2}{3}$	**d**	$\frac{3}{5}$ ☐ $\frac{1}{2}$
e	$\frac{5}{8}$ ☐ $\frac{7}{10}$	**f**	$\frac{1}{3}$ ☐ $\frac{2}{7}$	**g**	$\frac{3}{4}$ ☐ $\frac{5}{8}$	**h**	$\frac{3}{10}$ ☐ $\frac{2}{7}$
i	$\frac{3}{4}$ ☐ $\frac{4}{5}$	**j**	$\frac{7}{9}$ ☐ $\frac{2}{3}$	**k**	$\frac{7}{8}$ ☐ $\frac{3}{4}$	**l**	$\frac{1}{2}$ ☐ $\frac{2}{3}$
m	$\frac{3}{8}$ ☐ $\frac{1}{2}$	**n**	$\frac{2}{3}$ ☐ $\frac{7}{10}$	**o**	$\frac{2}{5}$ ☐ $\frac{3}{10}$	**p**	$\frac{5}{8}$ ☐ $\frac{2}{3}$
q	$\frac{4}{5}$ ☐ $\frac{8}{9}$	**r**	$\frac{7}{15}$ ☐ $\frac{1}{2}$	**s**	$\frac{11}{14}$ ☐ $\frac{6}{7}$	**t**	$\frac{6}{7}$ ☐ $\frac{7}{8}$

2 Use the $<$ and $>$ signs to make each statement correct.

a	$\frac{1}{3}$ ☐ $\frac{3}{8}$ ☐ $\frac{1}{4}$ ☐ $\frac{2}{5}$ ☐ $\frac{7}{10}$	**b**	$\frac{2}{3}$ ☐ $\frac{8}{10}$ ☐ $\frac{3}{4}$ ☐ $\frac{3}{5}$ ☐ $\frac{10}{15}$
c	$\frac{7}{9}$ ☐ $\frac{12}{16}$ ☐ $\frac{9}{18}$ ☐ $\frac{4}{10}$ ☐ $\frac{10}{20}$	**d**	$\frac{3}{18}$ ☐ $\frac{2}{10}$ ☐ $\frac{4}{16}$ ☐ $\frac{2}{9}$ ☐ $\frac{20}{100}$
e	$\frac{5}{8}$ ☐ $\frac{11}{22}$ ☐ $\frac{8}{20}$ ☐ $\frac{6}{12}$ ☐ $\frac{12}{18}$	**f**	$\frac{4}{8}$ ☐ $\frac{18}{27}$ ☐ $\frac{48}{100}$ ☐ $\frac{9}{12}$ ☐ $\frac{5}{6}$
g	$\frac{5}{6}$ ☐ $\frac{12}{16}$ ☐ $\frac{10}{15}$ ☐ $\frac{6}{10}$ ☐ $\frac{9}{12}$	**h**	$\frac{16}{20}$ ☐ $\frac{6}{8}$ ☐ $\frac{12}{18}$ ☐ $\frac{23}{50}$ ☐ $\frac{14}{21}$
i	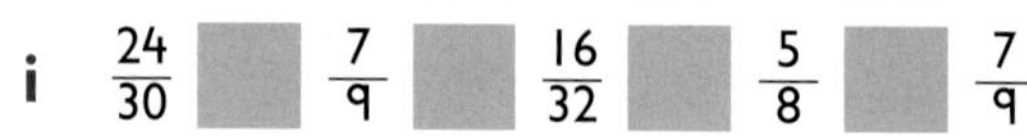 $\frac{24}{30}$ ☐ $\frac{7}{9}$ ☐ $\frac{16}{32}$ ☐ $\frac{5}{8}$ ☐ $\frac{7}{9}$	**j**	$\frac{14}{18}$ ☐ $\frac{15}{21}$ ☐ $\frac{18}{27}$ ☐ $\frac{12}{15}$ ☐ $\frac{16}{24}$

3 Order each set of fractions, smallest to largest.

a	$\frac{4}{9}$ $\frac{3}{9}$ $\frac{1}{9}$ $\frac{2}{3}$ $\frac{8}{9}$	**b**	$\frac{1}{3}$ $\frac{3}{4}$ $\frac{5}{6}$ $\frac{7}{12}$ $\frac{1}{6}$
c	$\frac{1}{2}$ $\frac{2}{5}$ $\frac{3}{4}$ $\frac{7}{10}$ $\frac{9}{10}$	**d**	$\frac{5}{6}$ $\frac{2}{3}$ $\frac{1}{3}$ $\frac{2}{9}$ $\frac{5}{12}$
e	$\frac{38}{100}$ $\frac{3}{5}$ $\frac{9}{25}$ $\frac{7}{10}$ $\frac{27}{50}$	**f**	$\frac{6}{15}$ $\frac{24}{30}$ $\frac{2}{18}$ $\frac{12}{20}$ $\frac{2}{6}$
g	$\frac{12}{18}$ $\frac{10}{16}$ $\frac{5}{10}$ $\frac{8}{20}$ $\frac{12}{27}$	**h**	$\frac{24}{36}$ $\frac{2}{12}$ $\frac{3}{6}$ $\frac{1}{3}$ $\frac{5}{18}$
i	$\frac{7}{9}$ $\frac{5}{8}$ $\frac{3}{4}$ $\frac{5}{6}$ $\frac{2}{3}$	**j**	$\frac{4}{5}$ $\frac{7}{8}$ $\frac{2}{3}$ $\frac{6}{7}$ $\frac{9}{10}$

Order the fractions

An activity to do by yourself or with a friend

You need:
- 2×0–9 dice

- Roll the dice to make a fraction, for example, a throw of 3 and 7 would make $\frac{3}{7}$.
- If you roll a 0 make it represent 10
- Write the fraction.
- Do this five times.
- Now order the fractions, smallest to largest.

Your teacher will tell you how many sets of fractions to order.

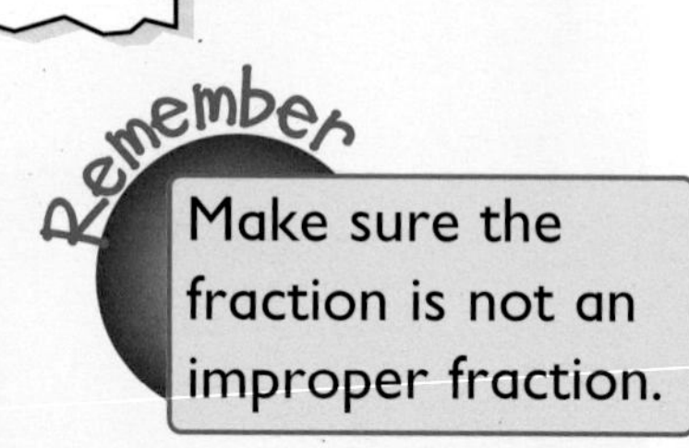

Make sure the fraction is not an improper fraction.

Adding and subtracting fractions

1 Choose 2 fractions from the football and add them together. Your teacher will tell you how many calculations to make.

Remember

- Change an improper fraction into a mixed number: $\frac{7}{5} = 1\frac{2}{5}$
- Reduce a fraction to its simplest form: $\frac{6}{8} = \frac{3}{4}$

2 Choose 3 fractions from the football and add them together. Your teacher will tell you how many calculations to make.

3 Choose 3 fractions from the football. Add two of the fractions together and subtract the third. Your teacher will tell you how many calculations to make.

4 Copy and complete.

a

+	$\frac{1}{5}$	$\frac{4}{5}$	$\frac{2}{5}$	$\frac{3}{5}$
$\frac{3}{5}$	$\frac{4}{5}$	$1\frac{2}{5}$		
$\frac{2}{5}$	$\frac{3}{5}$			
$\frac{1}{5}$				
$\frac{4}{5}$				

b

+	$\frac{4}{7}$	$\frac{1}{7}$	$\frac{3}{7}$	$\frac{5}{7}$
$\frac{2}{7}$				
$\frac{6}{7}$				
$\frac{3}{7}$				
$\frac{5}{7}$				

c

−	$\frac{2}{8}$	$\frac{1}{8}$	$\frac{4}{8}$	$\frac{3}{8}$
$\frac{7}{8}$				
$\frac{6}{8}$				
$\frac{4}{8}$				
$\frac{5}{8}$				

d

−	$\frac{3}{10}$	$\frac{4}{10}$	$\frac{5}{10}$	$\frac{2}{10}$
$\frac{7}{10}$				
$\frac{9}{10}$				
$\frac{8}{10}$				
$\frac{6}{10}$				

5 Copy and complete.

a

+	$\frac{1}{6}$		$\frac{5}{6}$	
	$\frac{5}{6}$			$1\frac{1}{6}$
$\frac{1}{6}$		$\frac{5}{6}$	1	
$\frac{3}{6}$	$\frac{2}{3}$			
			$1\frac{1}{6}$	

b

−		$\frac{4}{9}$		$\frac{5}{9}$
$\frac{7}{9}$	$\frac{5}{9}$		$\frac{4}{9}$	
$\frac{6}{9}$	$\frac{4}{9}$			
$\frac{5}{9}$				0
		$\frac{4}{9}$	$\frac{5}{9}$	

Multiplying fractions 1

1 Copy and complete.

a $\frac{2}{7}$ of 28 = ☐	**b** $\frac{4}{5} \times 20 =$ ☐	**c** $15 \times \frac{2}{3} =$ ☐	**d** $12 \times \frac{3}{4} =$ ☐
e $\frac{2}{5} \times 35 =$ ☐	**f** $\frac{5}{6}$ of 18 = ☐	**g** $\frac{3}{7} \times 42 =$ ☐	**h** $\frac{5}{9}$ of 45 = ☐
i $30 \times \frac{7}{10} =$ ☐	**j** $54 \times \frac{4}{9} =$ ☐	**k** $\frac{4}{7}$ of 21 = ☐	**l** $\frac{3}{5} \times 25 =$ ☐
m $40 \times \frac{5}{8} =$ ☐	**n** $\frac{5}{7} \times 49 =$ ☐	**o** $\frac{3}{10}$ of 70 = ☐	**p** $63 \times \frac{2}{9} =$ ☐
q $\frac{6}{7}$ of 35 = ☐	**r** $18 \times \frac{7}{9} =$ ☐	**s** $\frac{7}{8} \times 24 =$ ☐	**t** $\frac{3}{8} \times 32 =$ ☐

2 Choose a fraction and a number and multiply them together. Your teacher will tell you how many calculations to make.

- Change an improper fraction into a mixed number: $\frac{7}{5} = 1\frac{2}{5}$
- Reduce a fraction to its simplest form: $\frac{6}{8} = \frac{3}{4}$

$\frac{2}{3}$	$\frac{4}{5}$	$\frac{5}{7}$
$\frac{7}{9}$	$\frac{3}{5}$	$\frac{3}{4}$
$\frac{5}{6}$	$\frac{3}{8}$	$\frac{9}{10}$

35	42	32
21	20	18
48	24	15
14	50	12
27	45	36

3 Write the missing fractions.

a ☐ × 42 = 24	**b** ☐ × 15 ≠ 6	**c** 36 × ☐ = 30	**d** 18 × ☐ = 12
e 16 × ☐ = 10	**f** ☐ × 63 = 49	**g** ☐ × 50 = 45	**h** 49 × ☐ = 14
i 54 × ☐ = 48	**j** ☐ × 48 = 18	**k** 28 × ☐ = 21	**l** ☐ × 15 = 12
m ☐ × 21 = 9	**n** 36 × ☐ = 16	**o** ☐ × 20 = 6	**p** ☐ × 27 = 15
q 10 × ☐ = 6	**r** 28 × ☐ = 20	**s** 45 × ☐ = 10	**t** ☐ × 40 = 35

Multiplying fractions 2

1 Copy and complete.

a $\frac{2}{3}$ of 150 m = ☐ m
b $\frac{7}{10} \times 1$ l = ☐ l
c $\frac{2}{3} \times 120$ g = ☐ g
d $\frac{3}{4}$ of 24 hr = ☐ hr
e $\frac{3}{8}$ of 160 ml = ☐ ml
f $\frac{1}{2}$ of 186 m = ☐ m
g $\frac{3}{7}$ of 49 cm = ☐ cm
h $\frac{4}{9} \times 900$ ml = ☐ ml
i $\frac{7}{12}$ of 60 min = ☐ min
j $\frac{3}{4}$ of 1 km = ☐ km
k $\frac{4}{5} \times 60$ min = ☐ min
l $\frac{5}{7} \times 560$ g = ☐ g
m $\frac{4}{9} \times 90$ cm = ☐ mm
n $\frac{5}{8}$ of 2 l = ☐ ml
o $\frac{5}{6} \times 3$ hr = ☐ min
p $\frac{3}{8} \times 2$ kg = ☐ g
q $\frac{2}{9} \times 3$ years = ☐ months
r $\frac{3}{4} \times 5$ km = ☐ m
s $\frac{3}{5} \times 65$ cl = ☐ ml
t $\frac{5}{12}$ of 3 hr = ☐ min

2 Copy and complete.

- Change an improper fraction into a mixed number: $\frac{7}{5} = 1\frac{2}{5}$
- Reduce a fraction to its simplest form: $\frac{6}{8} = \frac{3}{4}$

a $\frac{2}{7}$ of 15 = ☐
b $\frac{4}{5} \times 19$ = ☐
c $\frac{2}{3} \times 20$ = ☐
d $15 \times \frac{3}{4}$ = ☐
e $\frac{3}{5} \times 10$ = ☐
f $19 \times \frac{5}{6}$ = ☐
g $\frac{2}{5}$ of 17 = ☐
h $\frac{4}{7} \times 9$ = ☐
i $6 \times \frac{3}{8}$ = ☐
j $\frac{4}{9}$ of 6 = ☐
k $\frac{2}{9}$ of 15 = ☐
l $\frac{5}{8} \times 20$ = ☐
m $\frac{3}{7}$ of 8 = ☐
n $\frac{8}{9} \times 10$ = ☐
o $13 \times \frac{5}{9}$ = ☐
p $\frac{3}{5}$ of 11 = ☐
q $7 \times \frac{7}{10}$ = ☐
r $\frac{5}{7} \times 100$ = ☐
s $\frac{7}{8} \times 12$ = ☐
t $\frac{4}{9} \times 7$ = ☐

The multiplying fractions game

A game for 2 players

- Shuffle the fraction cards and place them face down in a pile.
- Take turns to choose a card and roll the dice, for example $\frac{2}{7}$ and 18.
- Multiply the two numbers together: $\frac{2}{7} \times 18 = \frac{36}{7} = 5\frac{1}{7}$.
- Your score for that round is the whole number, i.e. 5.
- The winner is the player with most points after 10 rounds.

You need:
- Fraction cards (Resource Copymaster 12)
- 1–20 dice

Fractions, decimals and percentages 1

1 Write each of these fractions as a decimal and as a percentage.
Do not use a calculator.

a	$\frac{1}{2}$	**b**	$\frac{1}{4}$	**c**	$\frac{1}{100}$	**d**	$\frac{1}{10}$
e	$\frac{3}{4}$	**f**	$\frac{1}{5}$	**g**	$\frac{2}{3}$	**h**	$\frac{1}{3}$
i	$\frac{7}{10}$	**j**	$\frac{27}{50}$	**k**	$\frac{9}{20}$	**l**	$\frac{4}{5}$
m	$\frac{2}{5}$	**n**	$\frac{12}{25}$	**o**	$\frac{72}{100}$	**p**	$\frac{18}{25}$
q	$\frac{36}{50}$	**r**	$\frac{11}{20}$	**s**	$\frac{32}{400}$	**t**	$\frac{56}{200}$

2 Use a calculator to work out each of these fractions as a decimal and as a percentage. Round each decimal to 2 decimal places and each percentage to the nearest 1%.

a	$\frac{2}{11}$	**b**	$\frac{6}{15}$	**c**	$\frac{5}{6}$	**d**	$\frac{7}{12}$
e	$\frac{4}{7}$	**f**	$\frac{5}{9}$	**g**	$\frac{2}{9}$	**h**	$\frac{7}{8}$
i	$\frac{5}{13}$	**j**	$\frac{5}{8}$	**k**	$\frac{8}{21}$	**l**	$\frac{3}{17}$
m	$\frac{15}{26}$	**n**	$\frac{11}{14}$	**o**	$\frac{13}{28}$	**p**	$\frac{16}{22}$
q	$\frac{19}{35}$	**r**	$\frac{27}{52}$	**s**	$\frac{6}{5}$	**t**	$\frac{9}{7}$

3 Match the fraction and the decimal that together total 1.

$\frac{9}{72}$	$\frac{9}{20}$
$\frac{7}{8}$	$\frac{24}{60}$
$\frac{18}{48}$	$\frac{18}{72}$
$\frac{9}{15}$	$\frac{16}{25}$
$\frac{48}{64}$	$\frac{27}{90}$

0·125	0·6
0·55	0·25
0·875	0·7
0·625	0·36
0·4	0·75

Converting fractions to decimals

A game for 2 players

- Take turns to roll the dice, for example, 5 and 8.
- Make a fraction with the smaller number as the numerator and the larger number as the denominator, i.e. $\frac{5}{8}$.
- Now use the calculator to convert the fraction into a decimal rounded to two decimal places, i.e. 0·63.
- This is your score for that round.
- After 10 rounds, each player adds up their score.
- The winner is the player with the larger score.

You need:
- 2 × 0–9 dice (per pair)
- paper and pencil (per pair)
- a calculator (per pair)

Fractions, decimals and percentages 2

1 Write each of these percentages as a fraction and as a decimal.
Remember to reduce each fraction to its simplest form.

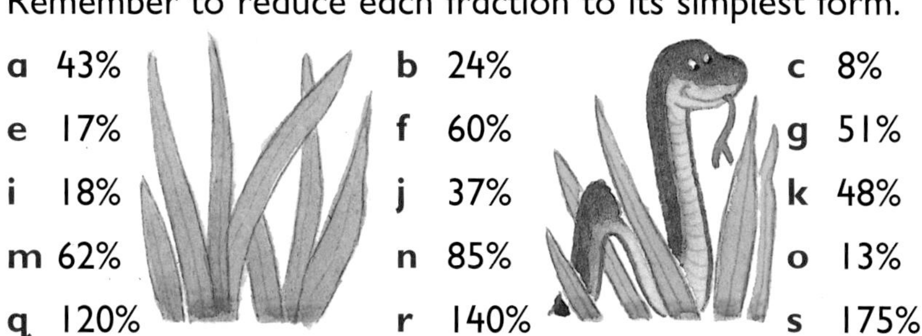

a 43%	b 24%	c 8%	d 32%
e 17%	f 60%	g 51%	h 72%
i 18%	j 37%	k 48%	l 26%
m 62%	n 85%	o 13%	p 99%
q 120%	r 140%	s 175%	t 280%

First to 100%

A game for 2 players

- Place your counters on 'Start'.
- Take turns to roll the dice and move your counter around the game board.
- As you keep moving around the game board add the fraction, decimal or percentage you land on to your previous total.
- The aim of the game is to be the first player to make 100% exactly. If a player makes more than 100% then they miss that turn and do not move their counter.

You need:
- a 1–6 dice (per pair)
- 2 counters, each of a different colour (per pair)

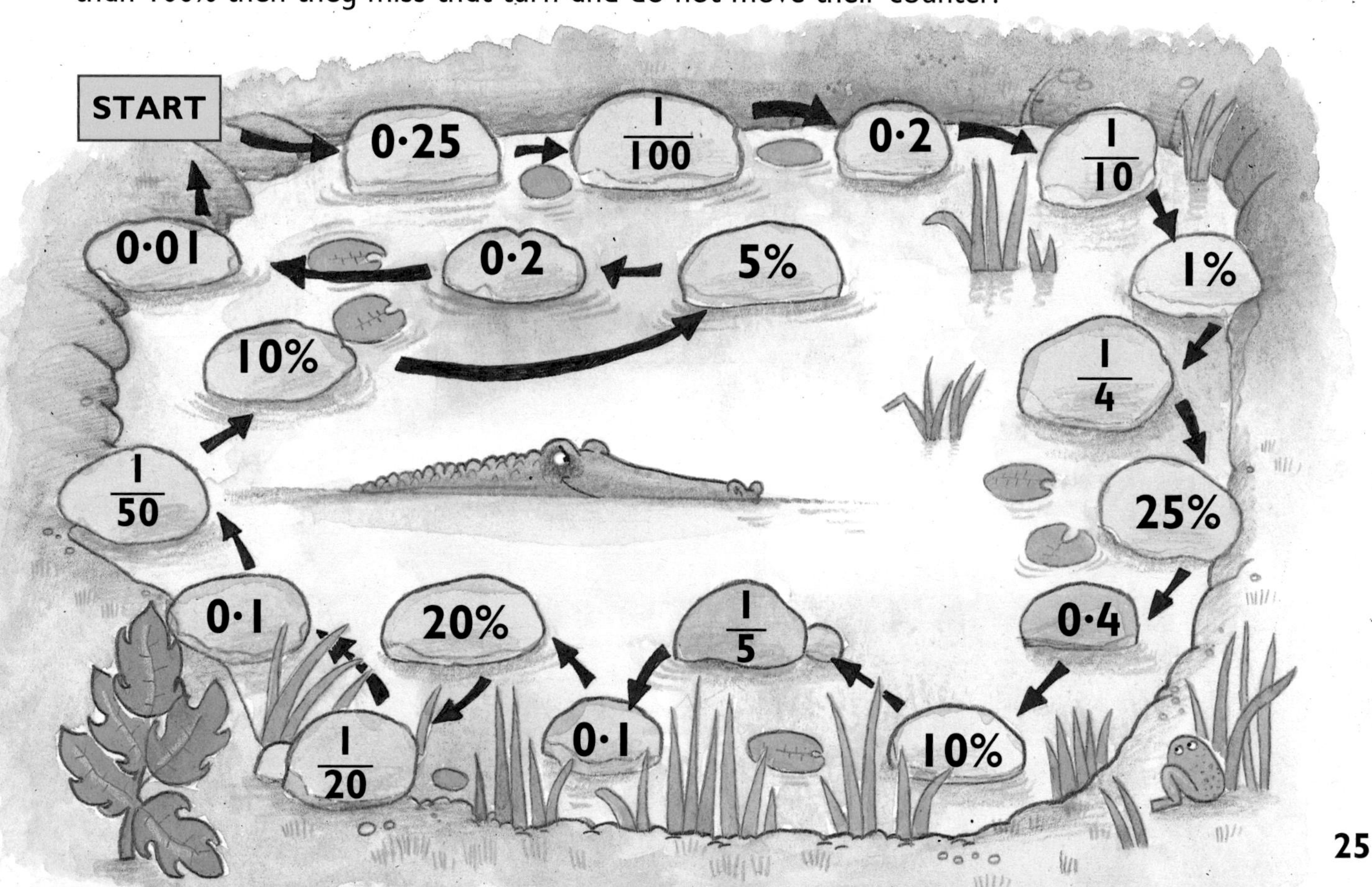

Calculating simple percentages

1 Estimate first, then calculate these percentages without using a calculator.

Example

26% of 5300 Estimate: 1300
10% = 530
5% = 265
1% = 53
$(530 \times 2) + 265 + 53 = 1378$

a 45% of 1800 **b** 16% of 7500 **c** 15% of 1600 **d** 36% of 2300
e 12% of 6200 **f** 26% of 2700 **g** 13% of 3500 **h** 89% of 5100
i 14% of 4800 **j** 18% of 3600 **k** 48% of 8200 **l** 21% of 4300
m 19% of 9300 **n** 22% of 3700 **o** 32% of 54 000 **p** 49% of 31 000
q 86% of 24 000 **r** 95% of 250 000 **s** 120% of 6800 **t** 149% of 51 000

2 Estimate first, then calculate the percentage of these amounts without using a calculator.

a 40% of 3200 ml **b** 61% of 5200 kg **c** 38% of £600 **d** 55% of 3400 km
e 99% of 1200 kg **f** 16% of 1200 cm **g** 80% of 3600 mm **h** 31% of 6200 l
i 54% of 3200 ml **j** 22% of 7100 g **k** 46% of 7800 cl **l** 65% of 9400 g
m 79% of 8 l **n** 92% of £4600 **o** 36% of 5 km **p** 84% of £70
q 65% of 9 m **r** 12% of £48 **s** 71% of 4 l **t** 24% of 7 kg

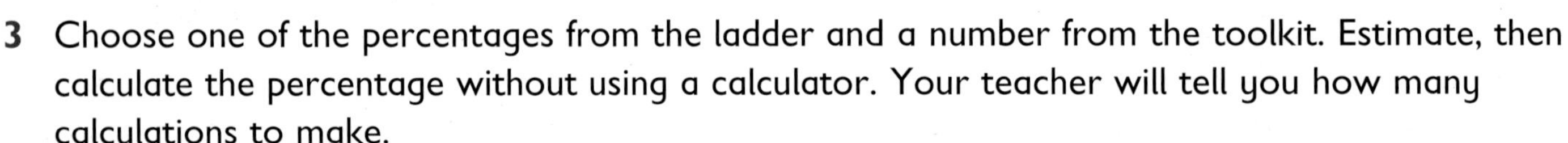

3 Choose one of the percentages from the ladder and a number from the toolkit. Estimate, then calculate the percentage without using a calculator. Your teacher will tell you how many calculations to make.

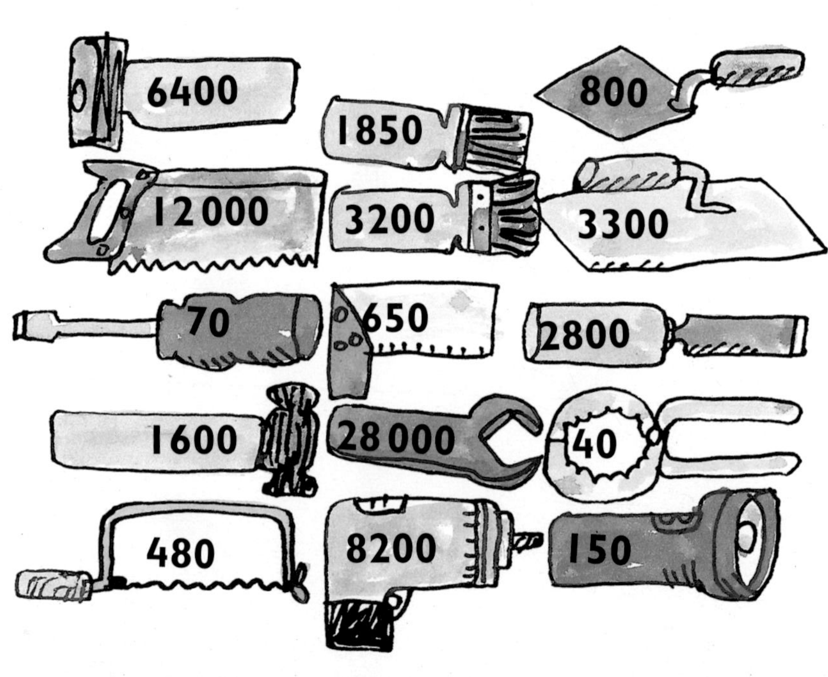

Calculating percentages

You need:
- a calculator

1 Estimate first, then calculate these percentages using a calculator.

a 13% of 3600	**b** 24% of 2500	**c** 17% of 2900	**d** 88% of 4300
e 42% of 8600	**f** 16% of 550	**g** 19% of 4600	**h** 37% of 3200
i 83% of 2400	**j** 92% of 1100	**k** 21% of 7900	**l** 39% of 500
m 57% of 5700	**n** 23% of 7100	**o** 34% of 12 000	**p** 48% of 96 000
q 14% of 41 000	**r** 18% of 360 000	**s** 148% of 69 000	**t** 221% of 570 000

2 Using a calculator, work out how much each price is reduced by, then write the new price for each item.

a £43, 15% off
b £68·20, 20% off
c £105, 5% off
d £78·70, 50% off
e £132, 12% off
f £217, 35% off

3 These prices are without VAT (value added tax). If VAT is 17·5%, work out the total cost for each item.

a £990
b £2700
c £8500
d £14 000
e £32 000
f £75 000

4 Answer these word problems.

a Marcus bought his flat in 1998 for £65 000. By 2002 its value had increased by a further 28%. What was the new value of the flat?

b The total attendance at last year's interschool football final was 8400. This year's attendance fell by 6%. How many people went to the final this year?

c Maurice has £50. He goes to the sports shop to buy a new football kit costing £45. When he gets to the shop, it is reduced by 15%. If Maurice buys the kit how much change will he get?

d Lee's food bill at a restaurant came to £67. On top of that the restaurant added a 12% service charge and Lee left another 10% of £67 as a tip. How much change did Lee get from £100?

e Last year the Davis family went on holiday to Nice. Their flights cost them £470 and their accommodation £840. This year the price of the flights has risen by 11% and the price of the accommodation has risen by 9%. What is the cost of their holiday this year?

f Pine trees are £32 each. However if you buy more than 5 there is a 30% reduction in the total price. If Lisa buys 12 trees how much will she have to pay?

Proportion

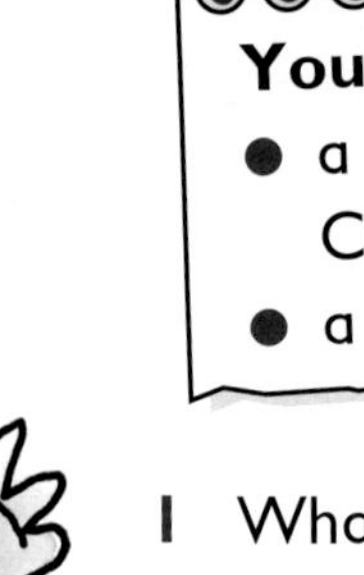

You need:
- a copy of Resource Copymaster 16
- a calculator

- Reduce a fraction to its simplest terms
- Round a decimal to two places
- Round a percentage to the nearest 1%

1 What proportion of the ingredients measured in grams in the Tomato Soup are:

a tomatoes? b onions?
c carrots?

Write each proportion as a fraction, a decimal and a percentage.

2 What proportion of the ingredients measured in millilitres or litres in the Tomato Soup are:

a olive oil? b tomato sauce?
c vegetable stock?

Write each proportion as a fraction, a decimal and a percentage.

3 What proportion of the ingredients measured in grams in the Mushroom Soup are:

a mushrooms? b onions?
c butter? d flour?

Write each proportion as a fraction, a decimal and a percentage.

4 What proportion of the ingredients measured in centilitres or litres in the Mushroom Soup are:

a milk? b vegetable stock?

Write each proportion as a fraction, a decimal and a percentage.

5 What proportion of the ingredients measured in grams in the Onion Soup are:

a onions? b butter?

Write each proportion as a fraction, a decimal and a percentage.

6 What proportion of the ingredients measured in grams in the Broccoli and Asparagus Soup are:

a broccoli? b asparagus?
c butter? d crème fraîche?

Write each proportion as a fraction, a decimal and a percentage.

Ratio

You need:
- a copy of Resource Copymaster 16

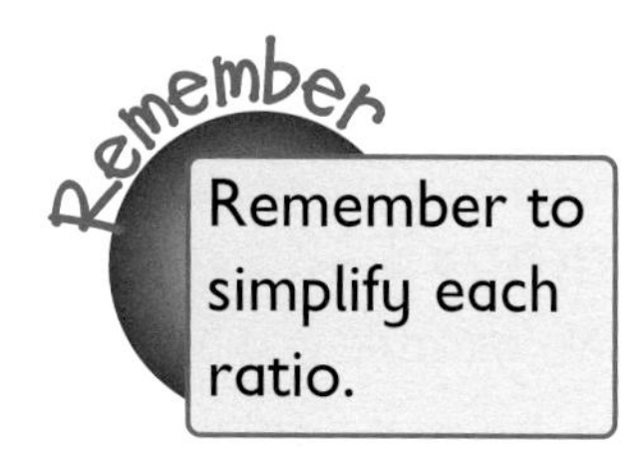

Remember to simplify each ratio.

1 For the Tomato Soup write the ratio of:

a tomatoes to carrots
b olive oil to tomato sauce
c olive oil to vegetable stock
d vegetable stock to tomato sauce

2 For the Mushroom Soup write the ratio of:

a mushrooms to butter
b butter to onions
c onions to mushrooms
d milk to vegetable stock

3 For the Onion Soup write the ratio of onions to butter.

4 For the Broccoli and Asparagus Soup write the ratio of:

a broccoli to asparagus
b broccoli to butter
c broccoli to crème fraîche
d asparagus to butter
e asparagus to crème fraîche
f crème fraîche to butter

5 Work out how much the chef has of each ingredient.

a

Ratio of asparagus to onions 5 : 7

b
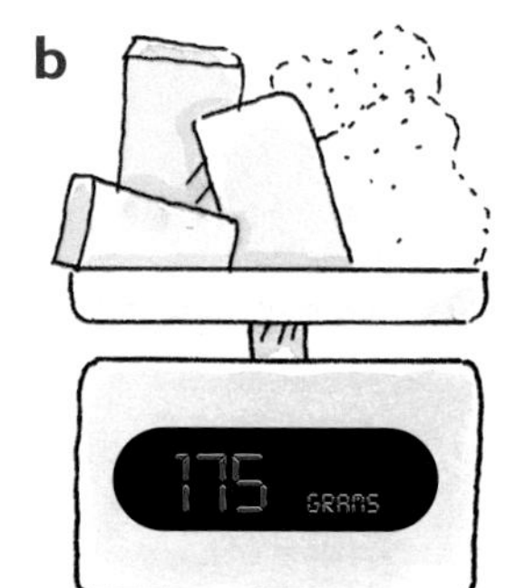

Ratio of butter to flour 3 : 4

c
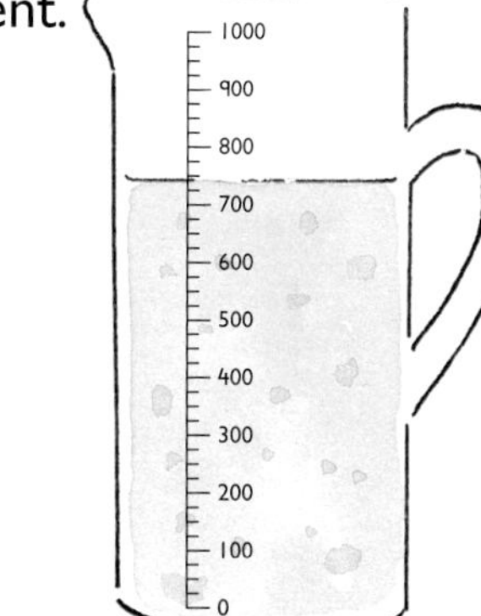

Ratio of white wine to vegetable stock 2 : 3

d

Ratio of apples to sugar 5 : 3

e

Ratio of rice to mushrooms 10 : 3

f

Ratio of asparagus to broccoli 9 : 7

g

Ratio of carrots to onions 3 : 5

h
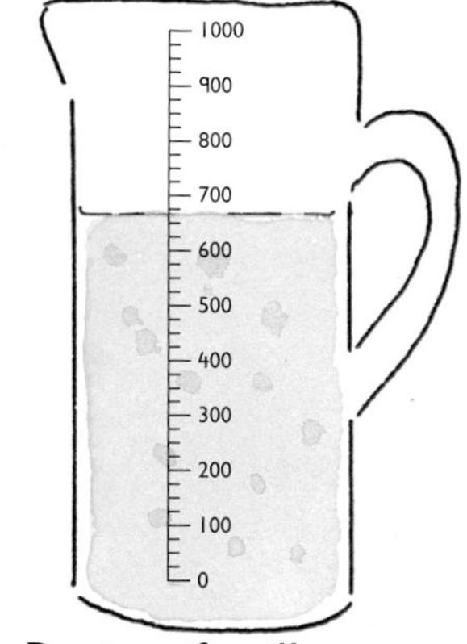

Ratio of milk to vegetable stock 2 : 7

Ratio and proportion

Remember
- Reduce a fraction to its simplest terms
- Round a percentage to the nearest 1%
- Round a decimal to two places
- Reduce a ratio to its simplest terms

Dice ratios and proportions

You need:
- a 1–6 dice (per pair)
- a calculator (per pair)

An activity for 2 players

- Take turns to roll the dice and record 10 numbers, for example: 4, 5, 2, 4, 1, 6, 2, 5, 6, 2.
- Now compare the numbers thrown and record the following:
  - **a** the proportion of even numbers;
  - **b** the proportion of numbers greater than 3;
  - **c** the ratio of even numbers to odd numbers;
  - **d** the ratio of ones to sixes;
  - **e** the ratio of twos to fours.
- Repeat the above 4 more times rolling the dice and recording 10 numbers each time.

Remember: Express proportions as a fraction, a decimal and a percentage.

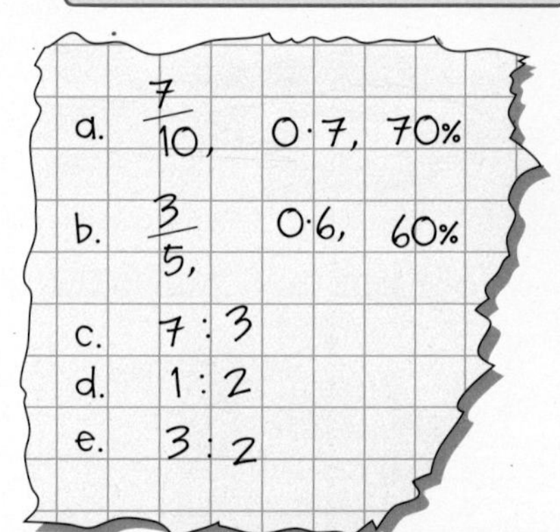

Variations:
- Try rolling the dice and recording 12, 15, 20 numbers at a time.
- Try using a 0–9 dice instead.

Card ratios and proportions

You need:
- a pack of playing cards (per pair)
- a calculator (per pair)

An activity for 2 players

- Shuffle the cards and place them face down in a pile.
- Take turns to turn over the top 8 cards and lay them out face up as a line on the table, for example:

    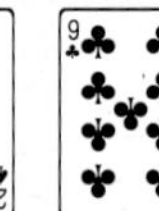   

- Now compare the cards and record the following:
  - **a** the proportion of black cards;
  - **b** the proportion of hearts;
  - **c** the ratio of black cards to red cards;
  - **d** the ratio of diamonds to spades;
  - **e** the ratio of hearts to diamonds to clubs.
- Repeat the above 4 more times taking 8 new cards each time.

Remember: Express proportions as a fraction, a decimal and a percentage.

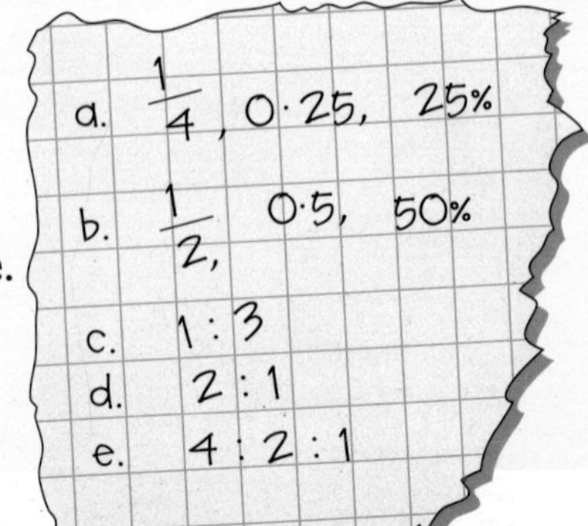

Variation:
- Try choosing 10, 12, 15 cards at a time.

Ratio and proportion problems

1 £1 is worth 1·58 Euros. On a recent trip to France, David bought a jumper costing 86 Euros. How much is it in pounds and pence?

2 A local supermarket is selling 12 oranges for £4·20. What is the cost of 7 oranges?

3 Paul brings a packet of chocolate biscuits to school. In the morning he shares half the packet with Jane and Peter. In the afternoon he shares the other half with Sally. If there are 24 biscuits in the packet, how many biscuits do each of them have?

4 Louis had £45 in birthday money. He spent some of the money on a CD. He saved four times as much as he spent. How much did he save?

5 A supermarket has 20 checkouts. There are 12 checkouts open. Write the ratio of open checkouts to closed checkouts in its simplest form.

6 Tommy and Gita play a game of marbles. At the end of the game Tommy has 20 marbles, $\frac{4}{5}$ of the number he started with, and Gita has 15 marbles.

   a How many marbles did Tommy start with?

   b What was the ratio of Tommy's marbles to Gita's marbles at the beginning of the game?

7 Kate and Abdul go to the cinema. Their tickets cost £4 each. Abdul only has £2.50, so Kate pays for the rest of the cost of his ticket, as well as her own. What proportion of the total cost does Kate pay?

8 £952 was collected for 3 charities – Save the Seals, Help the Budgies and Rabbit World. The money is to be divided in the ratio 3 : 4 : 7. How much does each charity receive?

9 In a cricket match only four players of one team get to bat. James scores 47, Terry 38, Jimmy 32 and Frank 8. What proportion of the total did each player score, expressed as a percentage?

10 I have a bottle each of lime juice, sugar syrup and soda water but only 45 cl of fruit purée. How much fruit cocktail can I make?

Fruit Cocktail

1 part lime juice
2 parts sugar syrup
5 parts fruit purée
4 parts soda water

Understanding multiplication and division

Remember Do not use a calculator

1 Multiply each of these numbers by 0, 1, 10, 100 and 1000.

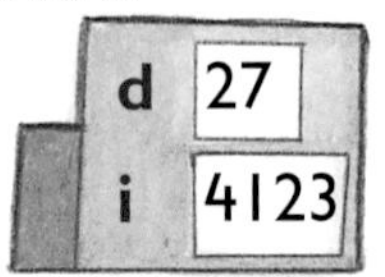
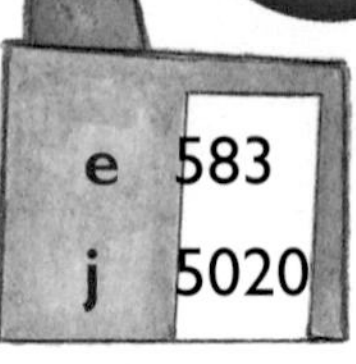

a	6	b	9	c	12	d	27	e	583
f	701	g	264	h	1542	i	4123	j	5020

2 Divide each of these numbers by 1, 10, 100 and 1000.

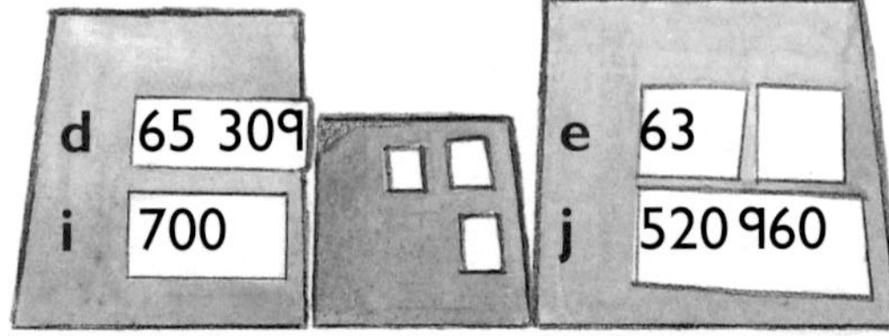

a	8	b	19	c	3854	d	65 309	e	63
f	1 849 032	g	455	h	1	i	700	j	520 960

3 Copy and complete. Write any remainders as a whole number remainder, as a fraction and as a decimal.

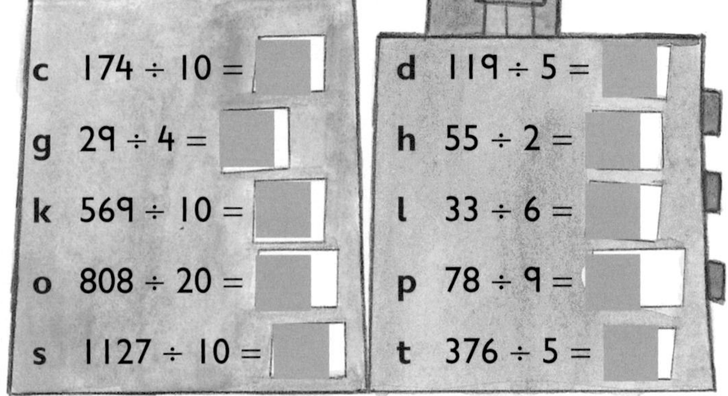

a $73 \div 2 =$	**b** $63 \div 5 =$	**c** $174 \div 10 =$	**d** $119 \div 5 =$
e $46 \div 4 =$	**f** $31 \div 3 =$	**g** $29 \div 4 =$	**h** $55 \div 2 =$
i $328 \div 20 =$	**j** $420 \div 100 =$	**k** $569 \div 10 =$	**l** $33 \div 6 =$
m $271 \div 100 =$	**n** $482 \div 8 =$	**o** $808 \div 20 =$	**p** $78 \div 9 =$
q $119 \div 2 =$	**r** $63 \div 4 =$	**s** $1127 \div 10 =$	**t** $376 \div 5 =$

4 Solve these word problems.

a Leroy has saved £25·50. For one day only Record World are selling all CDs for £6. How many CDs can Leroy buy?

b The gardeners at Park Madrid have to plant 8 flowerbeds. Each flowerbed must contain the same number of flowers. If there are 275 flowers, how many flowers will be in each flowerbed?

c On an aeroplane, each row of seats holds 9 passengers. How many rows will 250 passengers fill?

d Stan has 312 parcels to deliver to the post office. His van can only take 20 parcels at a time. How many trips to the post office will Stan have to make?

e The Empire State Building is 381 m tall. If each floor is 3 m high, how many floors are there in the Empire State Building?

f Francis the baker has had an order for 200 croissants from a local hotel. Each of his baking trays holds 12 croissants. How many trays does Francis use to bake all 200 croissants at the same time?

Inverse operations

1 Choose any two numbers from the balloon and add them together. Be sure to use a mental method with jottings. Show all your working. Now check your work. Show all your working again. Your teacher will tell you how many calculations to make.

2 Choose any two numbers from the balloon and multiply them together. Be sure to use a mental method with jottings. Show all your working. Now check your work. Show all your working again. Your teacher will tell you how many calculations to make.

Example

46 × 27
= (46 × 20) + (46 × 7)
= 920 + 322
= 1242

46 × 27
= (40 × 27) + (6 × 27)
= 1080 + 162
= 1242

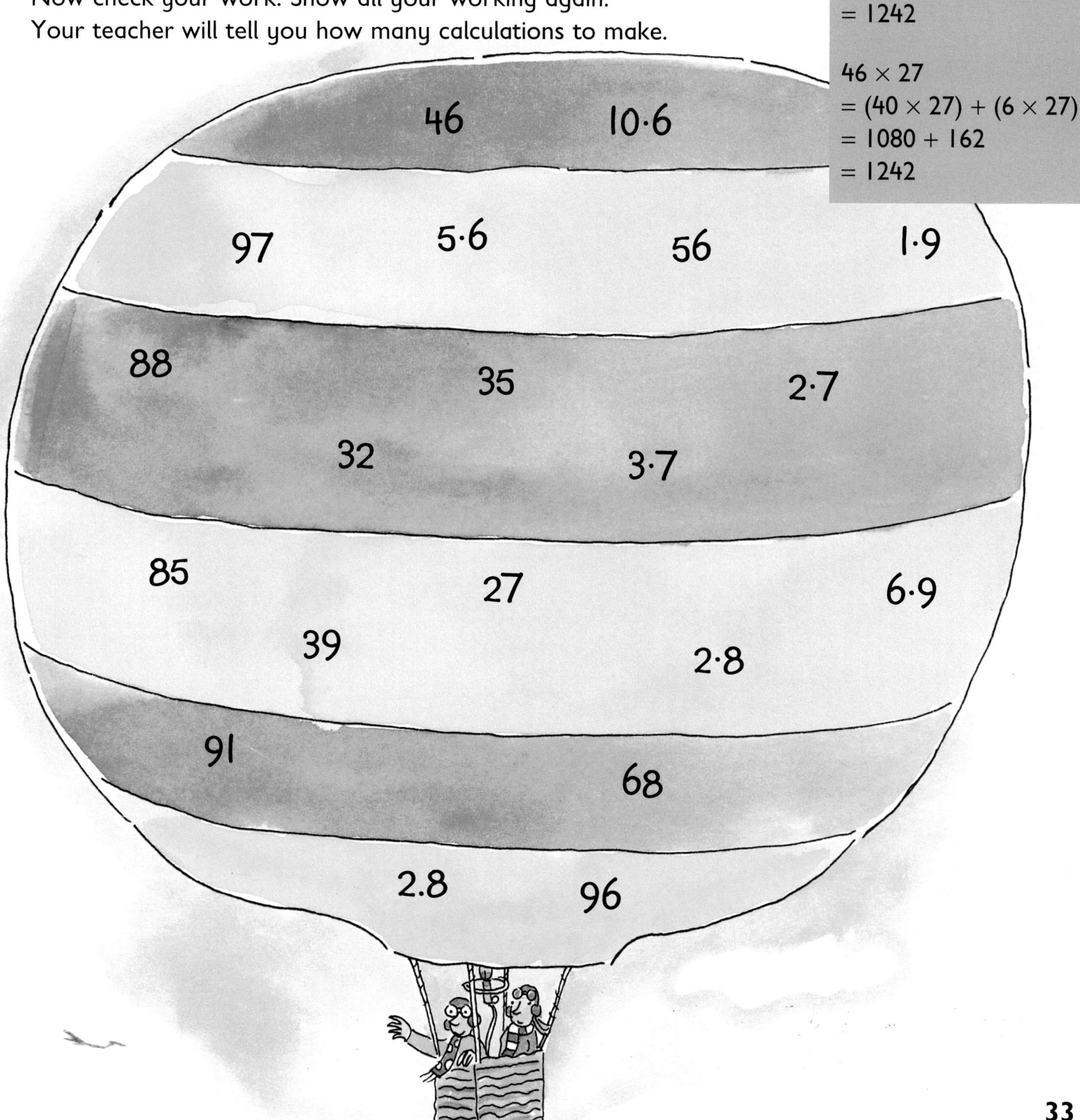

- Know and use the order of operations, including brackets
- Present and interpret solutions in the context of the original problem …
- Suggest extensions to problems by asking 'What if …?' …

Order of operations 1

1 Copy and complete.

a $15 + 9 \times 3 = \square$

b $43 - 27 \div 3 = \square$

c $9 \times 8 + 15 = \square$

d $54 \div 6 - 5 = \square$

e $5 \times 6 - 14 = \square$

f $56 - 3 \times 7 = \square$

g $23 + 63 \div 7 = \square$

h $89 - 88 \div 11 = \square$

i $12 - 15 \div 3 + 9 = \square$

j $15 + 8 \times 5 - 21 = \square$

k $54 \div 9 + 32 - 17 = \square$

l $42 - 3 \times 5 + 7 = \square$

m $16 + 81 \div 9 + 36 = \square$

n $87 - 55 + 6 \times 8 = \square$

o $6 \times 7 + 2 \times 21 = \square$

p $18 \div 6 + 4 \times 7 = \square$

q $9 \times 7 - 75 \div 5 = \square$

r $120 \div 10 + 17 \times 2 = \square$

s $9 \times 7 \div 3 + 39 = \square$

t $53 - 7 \times 54 \div 9 = \square$

2 Copy and complete.

a $(7 \times 4) - 25 = \square$

b $35 + (8 \times 6) = \square$

c $(49 \div 7) + 66 = \square$

d $99 - (9 \times 9) = \square$

e $98 - (64 \div 2) = \square$

f $(13 \times 2) + (25 \div 5) = \square$

g $(85 \div 5) - (72 \div 8) = \square$

h $(6 \times 9) + (6 \times 7) = \square$

i $(25 \times 5) - (48 \times 2) = \square$

j $(23 + 32) \div (32 - 27) = \square$

k $(57 - 33) \times (73 - 69) = \square$

l $(93 + 37) \div (43 - 38) = \square$

m $(87 - 73) \times (97 - 87) = \square$

n $(8 \times 7) \div (43 - 15) = \square$

o $(43 + 38) \div (27 \div 3) = \square$

p $(23 \times 4) \div (58 - 7 \times 8) = \square$

q $(72 \div 8) \times (54 \div 9) = \square$

r $(76 \div 2 + 8) - (7 \times 7 - 3) = \square$

s $(7 \times 8 - 23) \div (9 \times 8 - 69) = \square$

t $(8 \times 56 \div 7) \div (16 \times 94 \div 47) = \square$

Using the digits 2, 4, 5, 7 and 9 only, make each of the numbers below.
Can you find different ways to make these numbers?

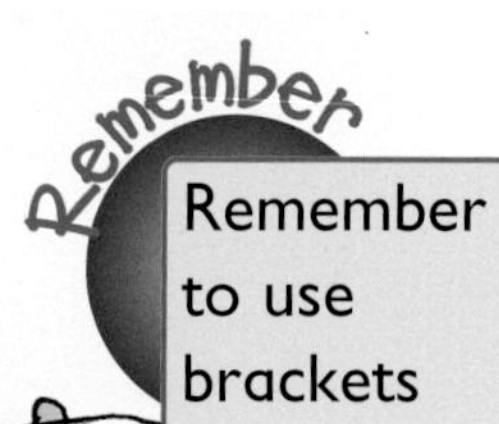

41 59 93 87 160

Order of operations 2

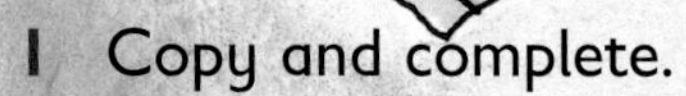

1 Copy and complete.

a $3^2 + (9 - 4) =$ ☐

b $7 \times (6 - 4)^2 =$ ☐

c $(19 - 13)^2 - 27 =$ ☐

d $4^2 + (3 \times 2)^2 =$ ☐

e $7^2 - (8 - 5)^2 =$ ☐

f $96 - (24 - 15)^2 =$ ☐

g $43 + (27 - 24)^2 =$ ☐

h $(34 - 28)^2 - 17 =$ ☐

i $(66 - 59)^2 + 38 =$ ☐

j $8^2 + (73 - 67)^2 =$ ☐

k $16 \div (5 + 3) =$ ☐

l $4^2 \div (12 - 10) =$ ☐

m $6^2 \div (7 - 5)^2 =$ ☐

n $8^2 \div (15 - 11)^2 =$ ☐

o $9^2 \div (2^2 + 5) =$ ☐

p $(4^2 - 9)^2 =$ ☐

q $(35 - 5^2)^2 =$ ☐

r $72 \div (7^2 - 43)^2 =$ ☐

s $(6^2 + 39) \div (54 - 7^2)^2 =$ ☐

t $(7^2 + 5^2) \div (6^2 + 1) =$ ☐

2 Match the calculation to the answer.

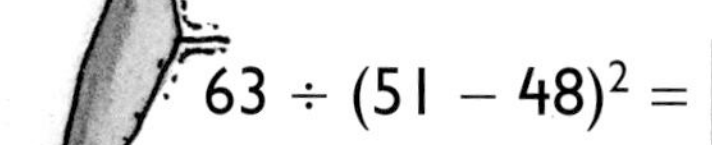

$5^2 - (3^2 - 2^2)^2 =$ ☐

$(9 + 7^2) \div (4^2 + 13) =$ ☐

$(3 + 5)^2 \div 2^2 =$ ☐

$(16 - 9)^2 + 5^2 =$ ☐

$(10^2 - 16) \div (4^2 - 3^2) =$ ☐

0   74   39   7   9   16   12   3   2   34

$63 \div (51 - 48)^2 =$ ☐

$3^2 \div (17 - 14) =$ ☐

$(1^2 + 4^2) \times 2 =$ ☐

$(29 - 23)^2 \div (3^2 - 5) =$ ☐

$(42 + 6^2) \div (5^2 - 23) =$ ☐

What's the calculation?

You need:
- a set of 0–9 digit cards (per pair)

A game for 2 players

- Shuffle the cards and place them face down in a pile.
- Turn over four cards, for example, 2, 4, 8, 7.
- If 2 is turned over it stands for the power of 2, i.e. ^{2}.
- Using all four digits each player makes a calculation and works out the answer, for example, $7^2 - (4 \times 8) = 17$.
- Keeping the calculation a secret, show the other player the answer.
- Each player now tries to make a calculation that gives that answer.
- Award points for each round as follows:
  - 1 point for being the first player to make your opponent's answer.
  - 2 points for being the first player to make your opponent's answer with the same calculation.
- The winner is the player with more points after 5 rounds.

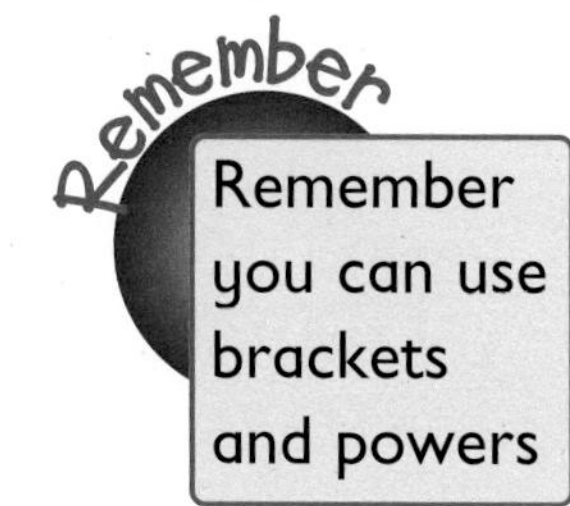

Remember you can use brackets and powers

Whole number addition and subtraction facts

1 Answer these addition and subtraction number facts to 20 using the information in the table.

□	○	□ + ○	□ − ○
14	5	19	9
11	7		
16	4		
10	8		
8	6		
20	0		
9	8		
6	5		
12	3		
17	2		

2 For each number card above, write the number that must be added to it, to make 100.
Write each complement as an addition calculation.

Example
36 + 64 = 100

3 For each number card above, write the number that must be added to it, to make 50.
Write each complement as an addition calculation.

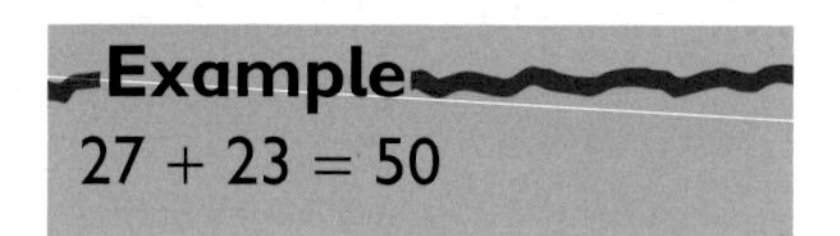

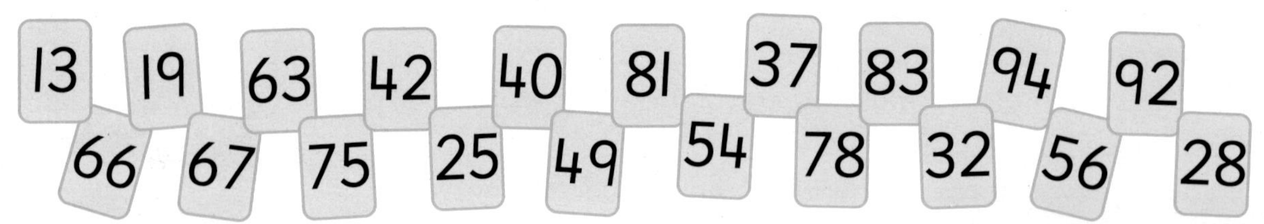

4 Choose two number cards from above. Add these two numbers together.
Your teacher will tell you how many calculations to make.

Example
42 + 83 = 125

5 Choose two number cards from above. Find the difference between these two numbers.
Your teacher will tell you how many calculations to make.

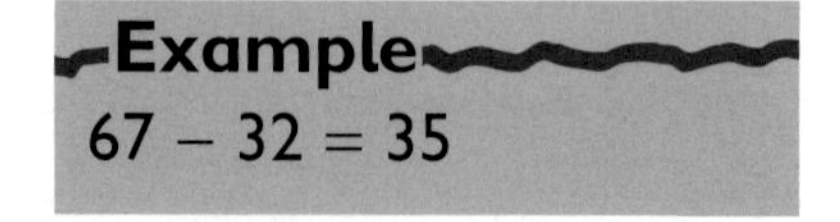

Decimal addition and subtraction facts

1 For each decimal number, write the decimal that must be added to it to make 1.
Write each complement as an addition calculation.

Example
0·37 + 0·63 = 1

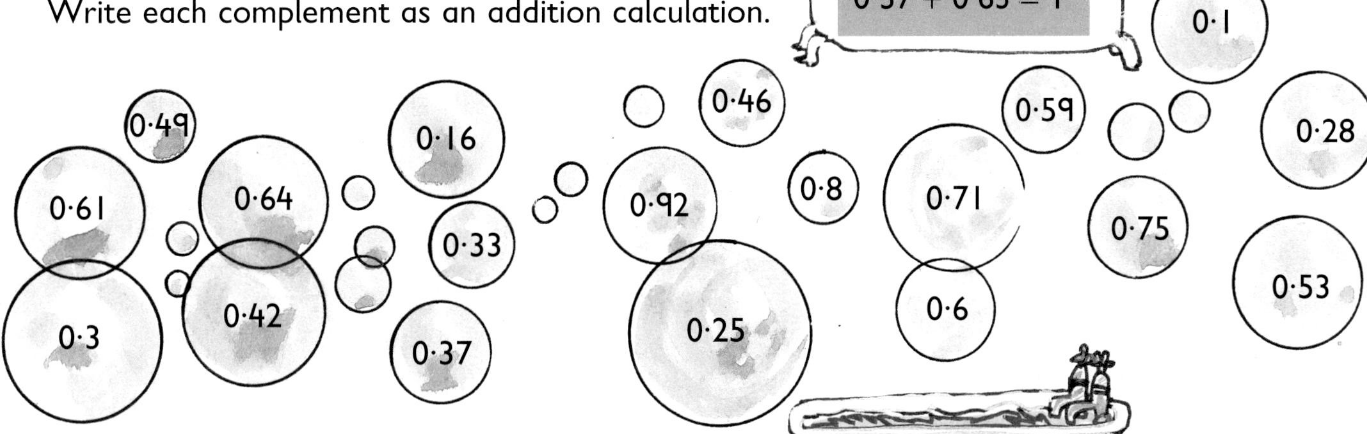

2 For each decimal number, write the decimal that must be added to it to make 10.
Write each complement as an addition calculation.

Example
1·87 + 8·13 = 10

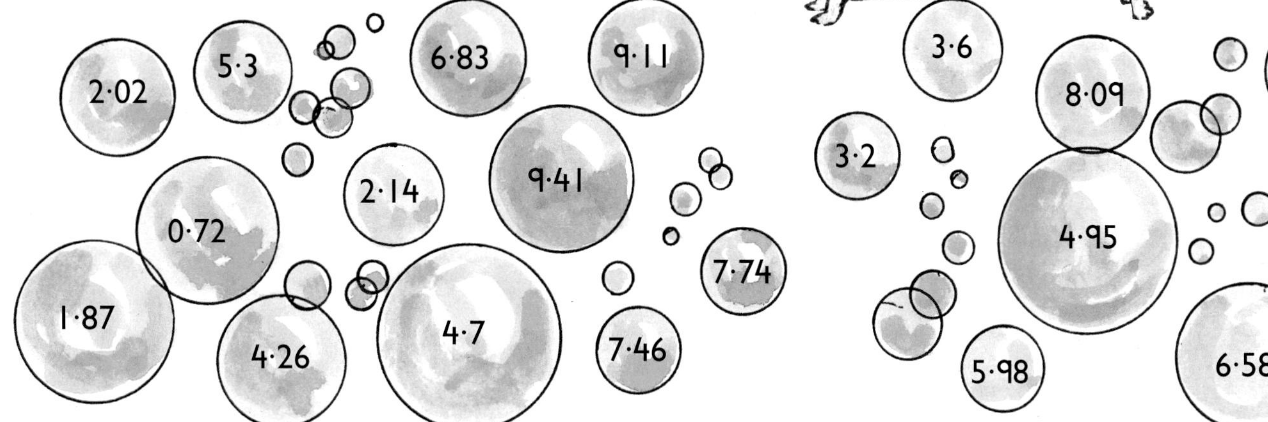

Add a tenth

You need:
- a calculator (per pair)
- pencil and paper (per pair)

A game for 2 players
- One player enters a decimal below 5 to one decimal place into the calculator, for example, 3·7.
- The other player has to add either 0·1, 0·2 or 0·3 to the number on the display.
- Keep taking turns to do this.
- The first player to reach 10 exactly is the winner of that round.
- Play 10 rounds. Who wins most rounds?

Variations:
- Start with 10 and subtract 0·1, 0·2 or 0·3. The first player to reach 0 exactly is the winner.
- Use hundredths. Enter a decimal below 3 to two decimal places, for example, 2·43 and take turns to add 0·11, 0·12 or 0·13. The first player to reach 10 exactly is the winner.
- Start with 10 and subtract 0·11, 0·12 or 0·13. The first player to reach 0 exactly is the winner.

- Consolidate the rapid recall of multiplication facts to 10×10 …
- Present and interpret solutions in the context of the original problem …
- Suggest extensions to problems by asking 'What if …?' …

Multiplication and division facts 1

Multiplication total

A game for 2 or more players

- Take turns to roll the dice, for example, 5 and 9 and make a multiplication fact.
- Write the calculation and the answer, i.e. $5 \times 9 = 45$
- Continue until each player has made 5 calculations.
- Each player then finds the total of the answers to their 5 calculations.
- The winner is the player with the larger total.
- Play the game again. This time the winner is the player with the smaller total.

You need:

- two 1–10 dice or a 0–9 dice with a 1 written in front of the 0 to represent 10 (per pair or group)
- pencil and paper (per child)

Division bingo

A game for 2 or more players

- Each player chooses 10 numbers from the grid and writes them down. These are your Bingo numbers.
- Take turns to roll the dice. If you roll a 1, have another turn.
- If one of the Bingo numbers can be divided exactly by the rolled number, cross out that number.
- The winner is the first player to cross out all of their Bingo numbers.

12	56	15	25	48
32	27	40	72	24
50	35	70	16	64
15	42	18	30	10
54	28	63	28	36
21	14	45	20	49

You need:

- 1–10 dice or a 0–9 dice with a 1 written in front of the 0 to represent 10 (per pair or group)
- pencil and paper (per child)

Puzzle time

a Arrange these numbers into pairs so that the difference between each pair is exactly divisible by 7.

85 88 58 66 86 43 18 38 72 37

b Help the burglar open the safe. The answers to the five puzzles in the code below give the combination to open it. Replace each symbol with a digit. Identical symbols must be replaced by the same digit and the same digit cannot be used for more than one symbol.

★ × ● = ■★ ◆ × ✖ = ✣✱ ✖ × ★ = ●★ ✖ × ● = ✱■ ◆ × ● = ■♥

Multiplication and division facts 2

Dice grids

4 activities to do with a friend

You need:
- a red 1–6 dice (per pair)
- a blue 1–6 dice (per pair)

For Grids 1 and 2:
- Take turns to roll the dice, for example, and find the matching number on the grid, i.e. 3400.
- Double the number, i.e. 6800.
- Write the answer as a calculation, i.e. 3400 × 2 = 6800.
- Your teacher will tell you how many calculations to make.
- If you roll the dice and get the same number as before, roll the dice again.

Grid 1

	⚀	⚁	⚂	⚃	⚄	⚅
⚀	62	5100	530	86	2700	670
⚁	120	59	270	8500	66	47
⚂	350	3400	73	490	5200	380
⚃	34	850	1300	23	51	3800
⚄	5300	6900	75	740	9200	660
⚅	780	48	1600	960	98	4900

Grid 2

	⚀	⚁	⚂	⚃	⚄	⚅
⚀	2·1	4·36	3·5	6·8	3·44	7·3
⚁	0·14	5·6	9·85	8·31	8·1	3·98
⚂	1·7	5·3	6·9	9·2	1·56	1·4
⚃	4·9	5·49	2·91	6·23	2·6	7·75
⚄	2·26	4·67	9·4	1·17	5·62	3·7
⚅	8·5	7·53	4·2	7·8	6·89	0·1

For Grids 3 and 4:
- Take turns to roll the dice, for example, and find the matching number on the grid, i.e. 17 400.
- Halve the number, i.e. 8700.
- Write the answer as a calculation, i.e. 17 400 ÷ 2 = 8700.
- Your teacher will tell you how many calculations to make.
- If you roll the dice and get the same number as before, roll the dice again.

Grid 3

	⚀	⚁	⚂	⚃	⚄	⚅
⚀	192	1410	106	174	1780	14300
⚁	13100	136	17400	1530	600	140
⚂	1250	800	1950	1370	88	1190
⚃	98	1320	18200	15700	126	19100
⚄	1830	164	12700	182	1680	16500
⚅	112	15200	1560	158	11900	1440

Grid 4

	⚀	⚁	⚂	⚃	⚄	⚅
⚀	0·2	0·42	1·02	3·4	6·16	9·28
⚁	0·38	0·26	4·6	5·62	8·56	6·2
⚂	5·4	8·6	3·48	1·8	0·92	0·64
⚃	2·76	0·94	0·32	0·82	8·24	1·78
⚄	2·8	1·2	0·58	0·18	4·2	5·72
⚅	0·06	7·4	4·34	0·76	7·54	9·6

Adding mentally

1 Work out the answers to these whole number addition calculations. Show your working.

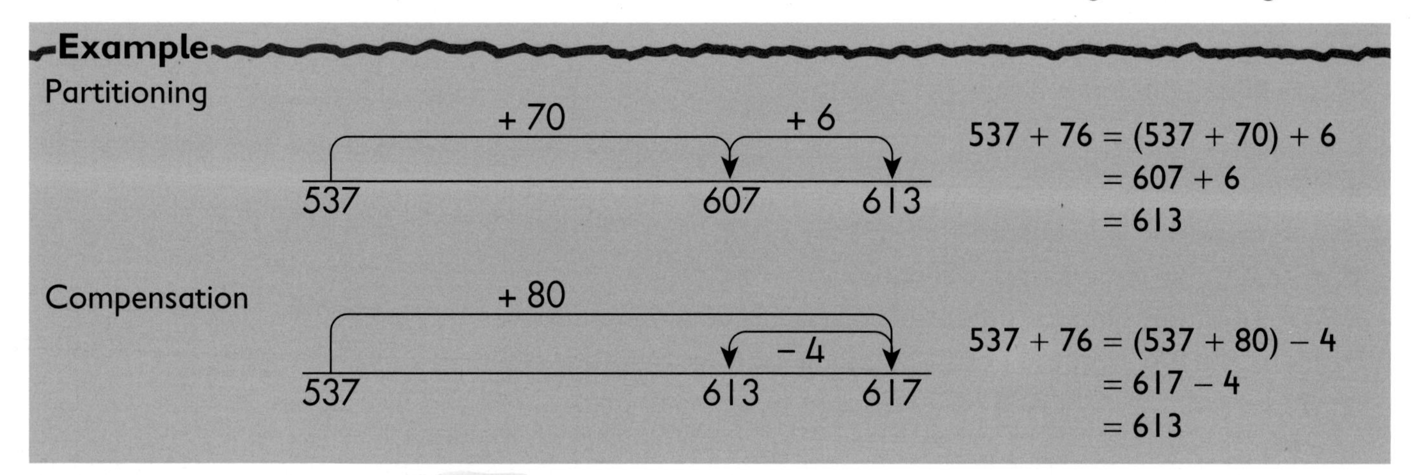

$537 + 76 = (537 + 70) + 6$
$= 607 + 6$
$= 613$

$537 + 76 = (537 + 80) - 4$
$= 617 - 4$
$= 613$

- **a** 385 + 67
- **b** 246 + 85
- **c** 437 + 58
- **d** 658 + 74
- **e** 586 + 35
- **f** 169 + 43
- **g** 853 + 38
- **h** 734 + 57
- **i** 596 + 69
- **j** 488 + 43

2 Now check your answers using a different method. Show your working.

3 Work out the answers to these decimal addition calculations. Show your working.

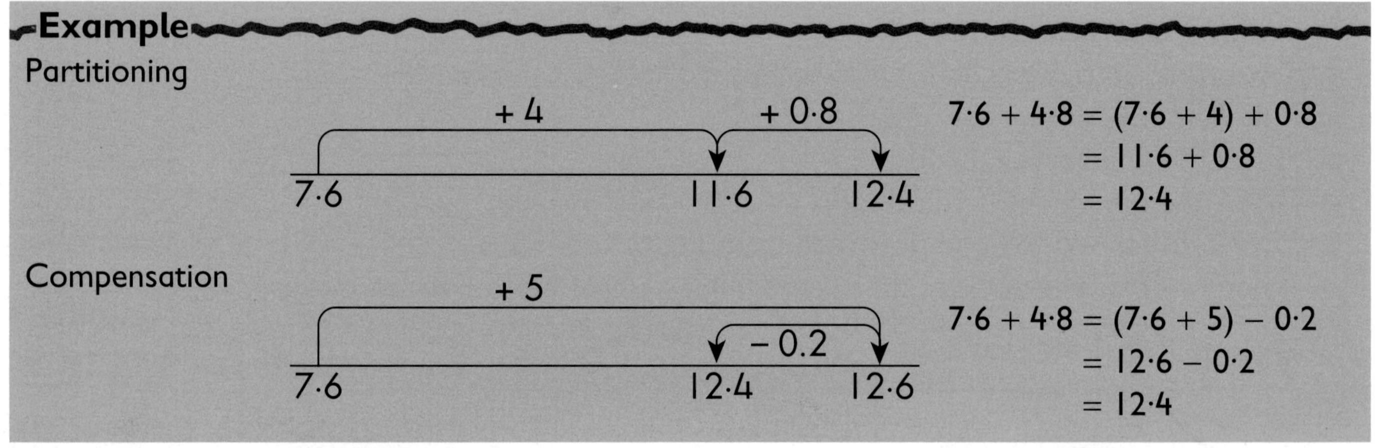

$7{\cdot}6 + 4{\cdot}8 = (7{\cdot}6 + 4) + 0{\cdot}8$
$= 11{\cdot}6 + 0{\cdot}8$
$= 12{\cdot}4$

$7{\cdot}6 + 4{\cdot}8 = (7{\cdot}6 + 5) - 0{\cdot}2$
$= 12{\cdot}6 - 0{\cdot}2$
$= 12{\cdot}4$

- **a** 8·7 + 4·6
- **b** 6·8 + 9·5
- **c** 12·4 + 7·7
- **d** 8·8 + 14·8
- **e** 9·4 + 13·9
- **f** 17·5 + 3·7
- **g** 26·6 + 5·8
- **h** 34·5 + 7·6
- **i** 47·8 + 5·5
- **j** 32·7 + 9·3

4 Now check your answers using a different method. Show your working.

Subtracting mentally

1 Work out the answers to these whole number subtraction calculations. Show your working.

Example

Partitioning

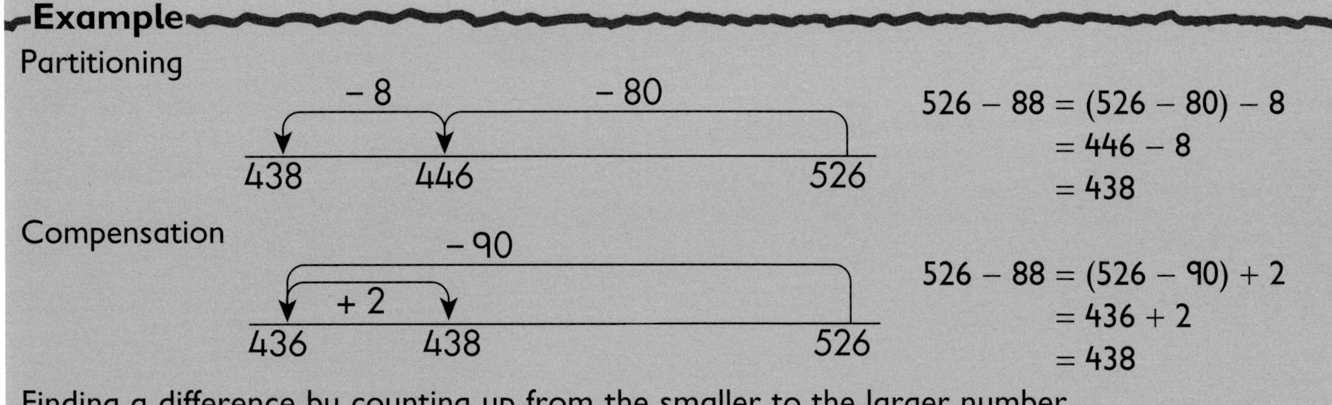

526 − 88 = (526 − 80) − 8
= 446 − 8
= 438

526 − 88 = (526 − 90) + 2
= 436 + 2
= 438

Finding a difference by counting up from the smaller to the larger number

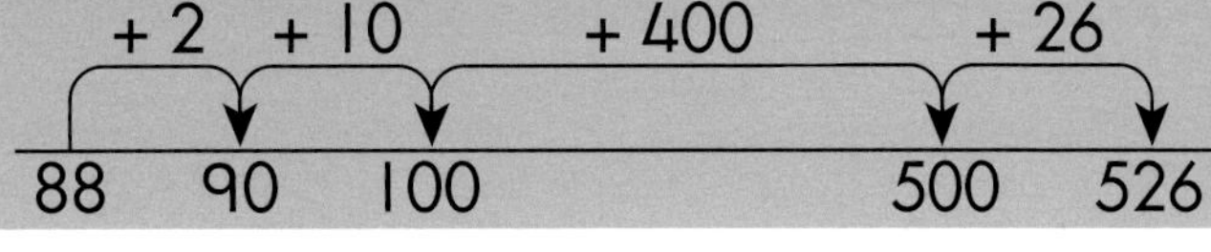

526 − 88 = 2 + 10 + 400 + 26
= 438

a 847 − 68 **b** 312 − 56 **c** 654 − 47 **d** 703 − 75

e 511 − 33 **f** 925 − 47 **g** 472 − 85 **h** 264 − 34

i 681 − 34 **j** 526 − 57

2 Now check your answers using a different method. Show your working.

3 Work out the answers to these decimal subtraction calculations. Show your working.

Example

Partitioning

− 0·6 − 7

8·7 9·3 16·3

16·3 − 7·6 = (16·3 − 7) − 0·6
= 9·3 − 0·6
= 8·7

Compensation

− 8 + 0·4

8·3 8·7 16·3

16·3 − 7·6 = (16·3 − 8) + 0·4
= 8·3 + 0·4
= 8·7

Finding a difference by counting up from the smaller to the larger number

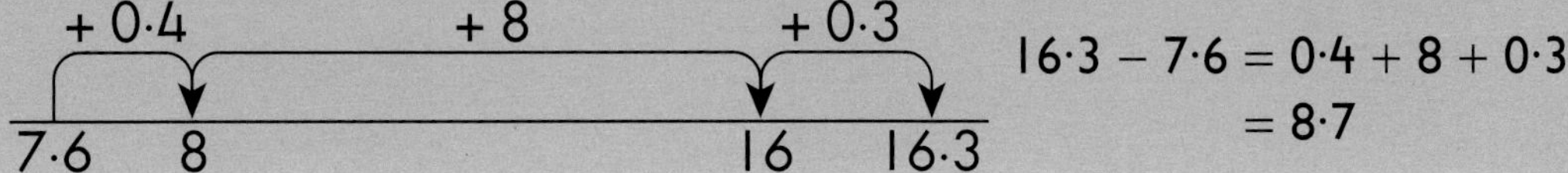

16·3 − 7·6 = 0·4 + 8 + 0·3
= 8·7

a 5·7 − 3·9 **b** 8·2 − 4·5 **c** 18·4 − 6·7 **d** 13·6 − 5·8

e 15·1 − 7·6 **f** 22·3 − 6·7 **g** 26·8 − 8·4 **h** 35·2 − 7·6

i 43·5 − 8·7 **j** 61·3 − 4·5

4 Now check your answers using a different method. Show your working.

Multiplying mentally

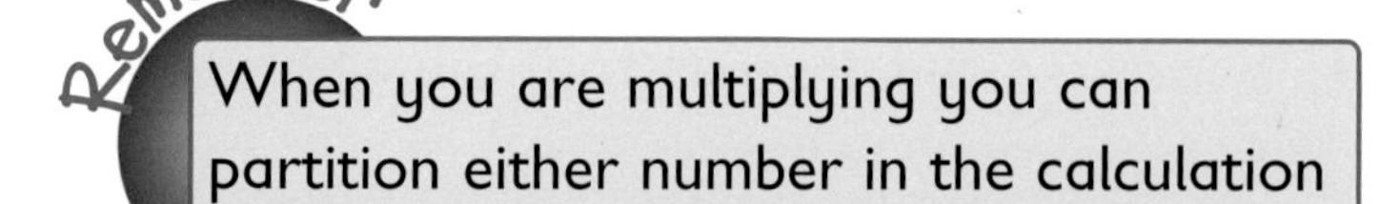

1 Work out the answers to these multiplication calculations. Show your working.

a	5.6×16	**b**	18×8.3
c	6.1×14	**d**	9.3×12
e	4.2×36	**f**	30×7.4
g	3.5×18	**h**	15×2.7
i	1.9×24	**j**	7.3×26
k	6.3×60	**l**	48×2.8

2 Now check your answers using a different method. Show your working.

3 Choose a number from the popcorn and a number from the fizzy drink so that when multiplied together they will make the following products.

Example

Factors

$$4{\cdot}6 \times 18 = (4{\cdot}6 \times 6) \times 3$$
$$= 27{\cdot}6 \times 3$$
$$= 82{\cdot}8$$

Partitioning

$$4{\cdot}6 \times 18 = (4{\cdot}6 \times 10) + (4{\cdot}6 \times 8)$$
$$= 46 + 36{\cdot}8$$
$$= 82{\cdot}8$$

or

$$4{\cdot}6 \times 18 = (4 \times 18) + (0{\cdot}6 \times 18)$$
$$= 72 + 10{\cdot}8$$
$$= 82{\cdot}8$$

Dividing mentally

Remember

Remember, when you are dividing you can only use factors of the number you are dividing by.

1 Work out the answers to these division calculations using factors. Show your working.

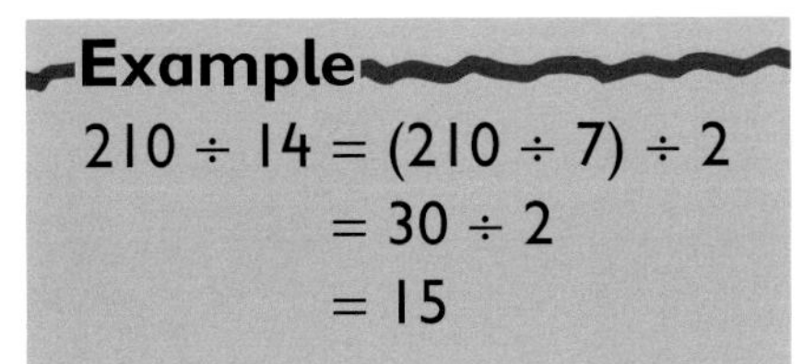

Example

$210 \div 14 = (210 \div 7) \div 2$
$= 30 \div 2$
$= 15$

a $336 \div 16$ **b** $350 \div 14$ **c** $465 \div 15$

d $276 \div 12$ **e** $342 \div 18$ **f** $144 \div 6$

g $490 \div 14$ **h** $384 \div 12$ **i** $432 \div 16$

j $525 \div 15$ **k** $396 \div 18$ **l** $320 \div 20$

2 Work out the answers to these division calculations using partitioning. Show your working.

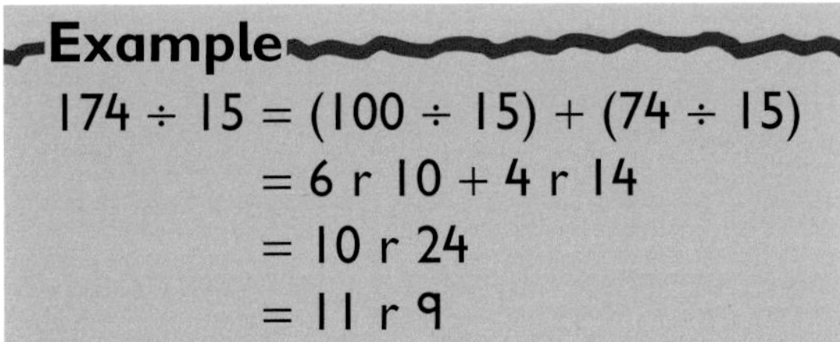

Example

$174 \div 15 = (100 \div 15) + (74 \div 15)$
$= 6 \text{ r } 10 + 4 \text{ r } 14$
$= 10 \text{ r } 24$
$= 11 \text{ r } 9$

Remember

Remember, when you are dividing you can only partition the number you are dividing into.

Mental problems and puzzles

1 When the last batsman at the Barton School cricket team went out to bat, the team's score was 229. If they needed another 37 runs to win, how many runs had their opponents scored?

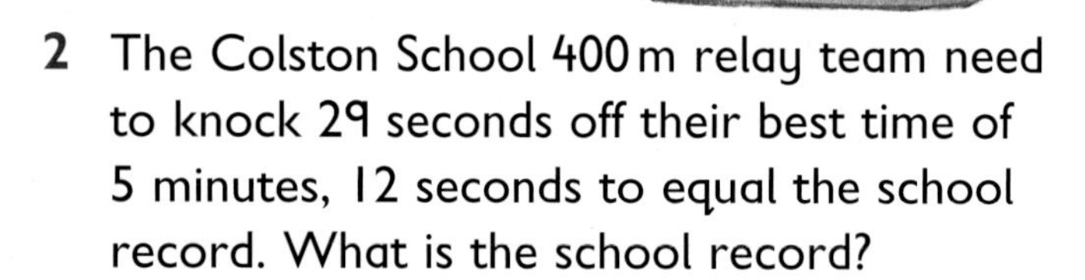

2 The Colston School 400 m relay team need to knock 29 seconds off their best time of 5 minutes, 12 seconds to equal the school record. What is the school record?

3 a Ahmed went shopping for a table tennis bat and some balls. The bat cost £8·90 and the box of balls cost £2·30. How much did he spend?

b After Ahmed had bought the bat and balls he had £17·20 left. He then met up with Frank and they went to the cinema which cost him £4·80. How much money did he have left?

4 On a school trip, all 27 children are given a packed lunch. Each lunch weighs 1·2 kg. What is the total weight of the packed lunches?

5 Alex has 462 cm of ribbon to wrap 14 Christmas presents. If each piece of ribbon is to be the same length, how long can each piece of ribbon be?

6 Sally went shopping with her mum for a pair of trainers. The ones she wanted cost £69. Her mum also bought a pair for herself for £66. Because they spent over £100 they got a 20% reduction. How much did the two pairs of trainers cost?

7 What number am I? When my double is added to 2·5 the result is 15·5.

8 Three consecutive integers add up to 81. What are they?

9 Choose from 1, 2, 3, 4 and 5 to place in the boxes. You cannot use a number more than once in any calculation.

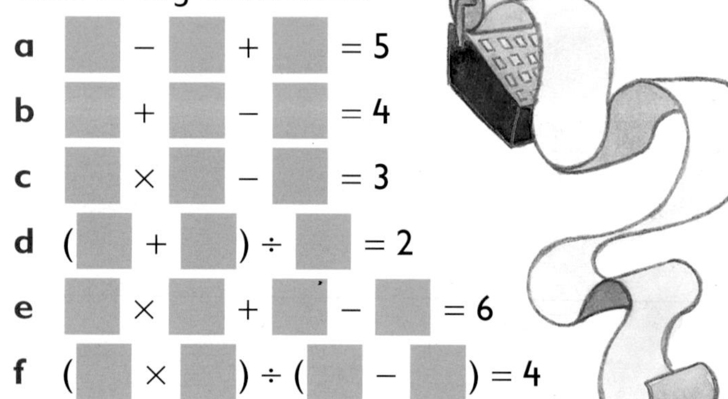

a ☐ − ☐ + ☐ = 5

b ☐ + ☐ − ☐ = 4

c ☐ × ☐ − ☐ = 3

d (☐ + ☐) ÷ ☐ = 2

e ☐ × ☐ + ☐ − ☐ = 6

f (☐ × ☐) ÷ (☐ − ☐) = 4

10 Write any number from 1 to 39. Multiply its last digit by 4 then add the other digit to this. Repeat the process until you get back to the original number. What is the longest chain you can make?

11 Complete these magic squares so that the sum of each row vertically, horizontally and diagonally is the same.

a

		5·5
4	6	
6·5		

b

3·75		4·75
	4	
		4·25

12 Each letter corresponds to a digit and the same letter represents the same digit. Write a number for each digit to make the calculation correct.

A B B × A = C D D

Adding whole numbers

Example

Approximate: 56 000

```
  49 132
     318
+  6259
  55 709
    111
```

1 Approximate first, then use a written method to work out the answers to these calculations.

a 5922 + 6943
b 389 + 529
c 53 029 + 62 302
d 74 512 + 18 029
e 1941 + 4941
f 487 + 926
g 4829 + 1241
h 243 198 + 192 484

2 Approximate first, then use a written method to work out the answers to these calculations.

a 841 + 9182
b 593 + 1949
c 19 481 + 491
d 76 + 4918
e 93 218 + 817
f 645 + 29 481
g 8419 + 10 491
h 11 476 + 6025

3 Approximate first, then use a written method to work out the answers to these calculations.

a 4819 + 59 + 91 842
b 91 + 716 + 9182
c 564 + 8391 + 91
d 4381 + 9182 + 48 + 371
e 5910 + 8173 + 69
f 58 + 9183 + 61 890 + 937
g 692 + 6502 + 69
h 591 + 952 + 69 206 + 194 204

4 Use the digits 1 to 9 to make the calculation correct. You can use each digit once only.

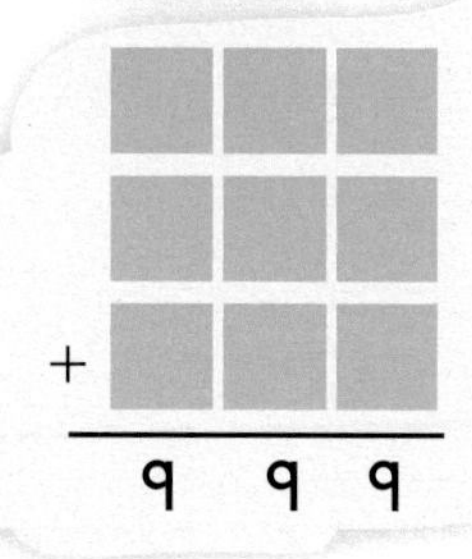

5 The letters A to J correspond to the digits 0–9. The same letter represents the same digit. Write a number for each digit to make the calculations correct. What do the letters A to J represent?

a
```
    A A A
+       B
  B C C C
```

b
```
    A G E F
+   B C D G
    B C E G H
```

c
```
    J A B G
+       B G
    I D H E
    D F G E
```

Adding decimals

Example

Approximate: 71

```
  43·63
+ 27·59
  71·22
   111
```

1 Approximate first, then use a written method to work out the answers to these calculations.

a 36·9 + 46·5 = ☐ **b** 76·4 + 18·8 = ☐

c 63·3 + 55·6 = ☐ **d** 87·67 + 69·73 = ☐

e 195·43 + 76·58 = ☐ **f** 368·7 + 259·4 = ☐

g 583·49 + 391·65 = ☐ **h** 637·42 + 789·73 = ☐

2 Approximate first, then use a written method to work out the answers to these calculations.

a 33·42 + 57·9 = ☐ **b** 86·3 + 38·57 = ☐ **c** 113·6 + 72·82 = ☐

d 247·59 + 76·8 = ☐ **e** 179·7 + 352·67 = ☐ **f** 251·3 + 675·89 = ☐

g 668·42 + 785·6 = ☐ **h** 981·73 + 867·58 = ☐

3 Approximate first, then use a written method to work out the answers to these calculations.

a 16·3 + 25·4 + 5·23 = ☐ **b** 45·63 + 13·2 + 28·47 = ☐

c 73·81 + 47·67 + 51·5 = ☐ **d** 67·4 + 71·98 + 96·8 = ☐

e 45·3 + 122·86 + 79·2 = ☐ **f** 126·38 + 257·1 + 64·88 = ☐

g 36·47 + 89·3 + 71·4 + 8·98 = ☐ **h** 364·87 + 49·2 + 73·58 + 87·49 = ☐

Add the cards

A game for 2–4 players

- Each player shuffles their cards and deals out six each.
- Then, using their decimal point cards, places their cards in this arrangement:

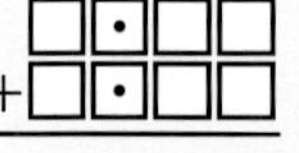

- Players then work out the answer to their calculation.
- The winner of each round is the player with the highest total, and that player collects a counter.
- The overall winner is the player with most counters after 10 rounds.

You need:

- two sets of 0–9 cards (per child)
- three decimal point cards (per child)
- pencil and paper (per child)
- 10 counters per pair or group

Variation:

- Change the arrangements of the cards, for example:

Subtracting whole numbers

Example

Approximate: 1500

```
  5 12
  6259
– 4715
  1544
```

1 Approximate first, then use a written method to work out the answers to these calculations.

a 4591 – 2013 =
b 5723 – 2948 =
c 481 – 214 =
d 12 945 – 10 489 =
e 582 – 129 =
f 9812 – 2819 =
g 19 848 – 13 044 =
h 1804 – 1092 =
i 129 804 – 105 492 =
j 503 903 – 280 094 =

2 Approximate first, then use a written method to work out the answers to these calculations.

a 8124 – 942 =
b 7455 – 498 =
c 12 953 – 849 =
d 16 071 – 4189 =
e 431 841 – 192 =
f 127 948 – 6298 =
g 81 481 – 2593 =
h 71 129 – 718 =
i 912 417 – 9814 =
j 481 941 – 84 712 =

3 Use the digits 1 to 9 to make the calculation correct. You can use each digit once only.

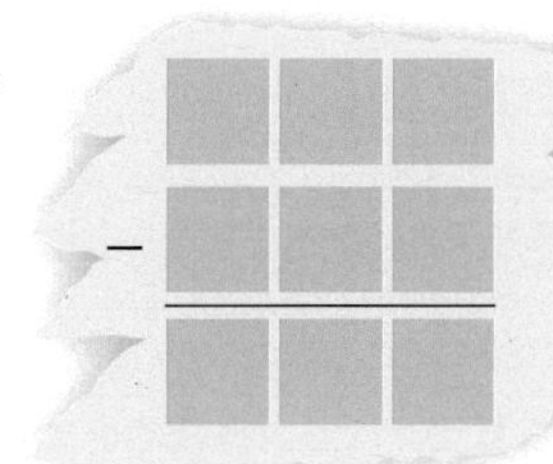

4 The letters A to J correspond to the digits 0–9. The same letter represents the same digit. Write a number for each digit to make the calculations correct. What do the letters A to J represent?

a
```
  A B C D E
–     F G H C
  F G D C I
```

b
```
  J G A E G
– I F G H J
      C D G J
```

5 A three-digit number is subtracted from a number made up of the same three digits in reverse order. The answer consists of the same three digits in yet another different order. What is the calculation?

Subtracting decimals

Example
Approximate: 16

$$\begin{array}{r} \overset{3}{\not4}\overset{13}{3}\cdot\overset{5}{\not6}\overset{13}{3} \\ -\ 27\cdot59 \\ \hline 16\cdot04 \end{array}$$

1 Approximate first, then use a written method to work out the answers to these calculations.

a 43·9 – 37·4 = ☐
b 58·3 – 13·7 = ☐
c 28·67 – 14·36 = ☐
d 76·42 – 31·77 = ☐
e 163·5 – 47·8 = ☐
f 64·37 – 28·96 = ☐
g 283·74 – 56·63 = ☐
h 332·4 – 83·7 = ☐
i 423·23 – 179·35 = ☐
j 385·88 – 73·79 = ☐

2 Approximate first, then use a written method to work out the answers to these calculations.

a 57·28 – 28·5 = ☐
b 46·29 – 12·6 = ☐
c 847·51 – 289·6 = ☐
d 402·3 – 175·37 = ☐
e 138·1 – 57·43 = ☐
f 28·52 – 12·7 = ☐
g 308·5 – 142·72 = ☐
h 58·03 – 15·26 = ☐
i 688·56 – 506·6 = ☐
j 479·2 – 192·08 = ☐

Subtract the cards

A game for 2–4 players

- Each player shuffles their cards and deals out six each.
- Then, using their decimal point cards, places their cards in this arrangement:

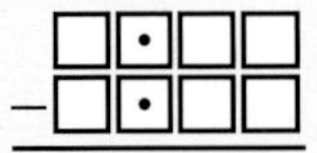

- Players then work out the answer to their calculation.
- The winner of each round is the player with the smallest answer, and that player collects a counter.
- The overall winner is the player with most counters after 10 rounds.

You need:
- two sets of 0–9 cards (per child)
- three decimal point cards (per child)
- pencil and paper (per child)
- 10 counters per pair or group

Variations
- Change the arrangements of the cards, for example:

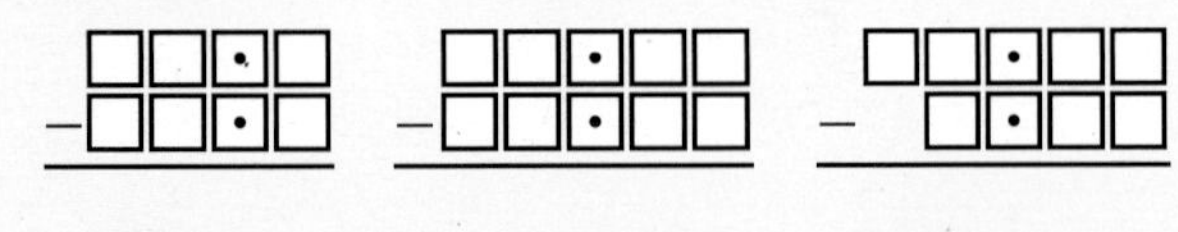

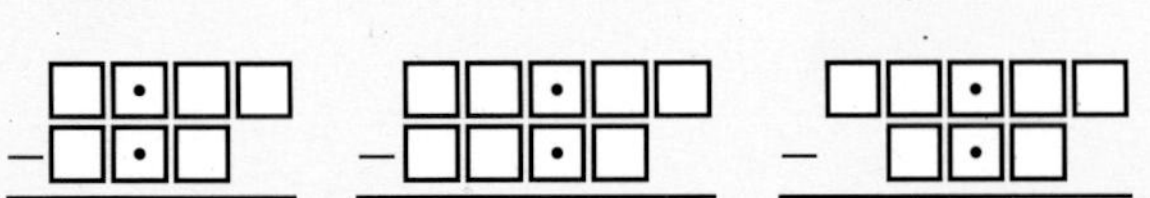

Addition and subtraction word problems

Solve these word problems.

1 The Ocean Princess cruise liner holds 2648 passengers and 387 crew. When the ship docks at Monte Carlo harbour, 1243 passengers and 109 crew get off for a visit. How many people are left on the ship?

2 The Martin Hotel has 197 single rooms, 538 double rooms, 104 family rooms and 18 suites. How many rooms are there in the hotel?

3 To make a kilo of strawberry ice-cream you mix 568 g of strawberries, 284 g of double cream, 29 g of lemon juice together with sugar and place it in the freezer, stirring it every half hour. How much sugar do you use?

4 56 549 people attended the Cannes Film Festival. Of these, 12 403 were producers and directors, 18 187 cameramen and technicians, 4297 journalists and 1689 film stars. The remainder were film fans. How many fans attended the festival?

5 The Pacific Ocean has an average depth of 3939 m. The Sea of Japan has an average depth of 1667 m. On average how much deeper is the Pacific Ocean than the Sea of Japan?

6 A local library has 32 648 fiction books and 27 574 non fiction titles. On the 17th June a total of 6248 books were out on loan. How many books were still in the library?

7 In 1912, 1517 people disappeared off the Titanic steamer. Three years later in 1915, 1198 people disappeared off the Lusitania steamer, and in 1917, 804 people disappeared off the battleship Vanguard. How many people altogether disappeared from the 3 ships?

8 A football stadium can hold 8410 people. It has four stands. Stand A holds 2468 people. Stand B holds 1612 people. Stand C holds 2537 people. How many people does Stand D hold?

9 The London Underground was opened in 1863 and is about 394 km in length. The New York Underground was opened in 1867 and is about 389 km in length. The Paris Underground was opened in 1900 and is about 199 km. What is the total length of all three Undergrounds?

10 CD World hope to sell 35 000 CDs this year. So far they have sold 11 298 rock CDs, 5083 classical CDs and 842 jazz CDs. How many more CDs do they need to sell before they reach their target?

11 Three airports service a large city. During one week, one airport dealt with 128 498 passengers, another with 97 581 and the third with 205 405. How many people did the three airports deal with during the week?

12 Tony and the Labourers did three concerts on their recent tour to London. On the Monday night they sold 3891 tickets, on the Wednesday night 5347 tickets, and on the Friday night 8212 tickets.

a What was the difference in the number of tickets they sold between the most and the least popular nights?

b If the hall in which they were performing could hold a maximum of 8500 people, how many more tickets could they have sold each night?

c How many more tickets could they have sold for all three nights?

More addition and subtraction word problems

1 Toni's garden is 42·60 m² in size. The garden shed takes up 5·75 m², the path 6·87 m², the flower beds 13·96 m² and the rest is all grass. How much grass is there for him to play on?

2 Brian the electrician is installing electric cables for a new kitchen. He has bought a reel of cabling 100 m long. Along one wall he uses 24·6 m of cable and along a second wall 27·8 m. How much cable does he have left?

3 Mr and Mrs Marfey have four children. Sarah is 12 years old and James is 14. Alec and Charlie are twins and are 8 years old. They are all going to visit Ocean World. A family ticket costs £64·50 and is for 2 adults and 2 children under the age of 10. Tickets are £18·75 each for children between the ages of 10 and 16. How much does it cost the Marfey family to visit Ocean World?

4 Misha's dad is building her three new shelves for her bedroom. One shelf is 2·35 m, another 3·5 m, and the third 4·65 m. He has bought 2 pieces of wood each 10 m in length. How much wood will he have left?

5 CD Warehouse is having a promotional offer – buy four CDs and get the cheapest one free. Sylvia buys four CDs which cost £11·74, £13·48, £9·52 and £12·63. How much does she pay?

6 Annie is taking part in a 2-day, 52·75 km cycle journey. On the first day, she cycled 11·54 km in the morning, 9·42 km in the afternoon and 6·3 km in the evening. How far does she have to travel on the second day?

7 The headquarters of a computer company are re-designing their lobby and have bought three new fully grown trees. One tree is 12·4 m high and cost £6341·46, another is 10·75 m high and cost £4208·67, the third is 8·2 metres and cost £8094·12. How much did the company spend on the three trees?

8 The lift in a department store can carry 575 kg maximum weight. Six people are waiting to enter the lift. Their weights are 48·7 kg, 72·6 kg, 54·2 kg, 91·6 kg, 67·3 kg and 59·75 kg. There are already two people in the lift, one person weighs 51·9 kg and the other 62·25 kg.

a What is the combined weight if all 8 people enter the lift?

b Can another person travel in the lift with these 8 people? What is the maximum weight that person can be?

9 There are 272·9 million people in the United States of America. Of these, 59·22 million are under 15 years of age and 43·94 million are over 65. How many people between the ages of 15 and 65 are there in the United States of America?

10 For sports day, a school made 80 litres of orange squash. By 10:30 the infants' school have used 9·7 litres and the junior school have used 22·25 litres. How much squash do they have left?

11 Mr Taylor had £8054·32 in his bank account. In one day he took out £100 from the cash machine and wrote three cheques. One cheque was for £259·23, another for £1026·87 and the third was for £86·14. How much money does Mr Taylor have left in his account?

12 Robbie the green-keeper wants to spray the grass at the local golf club. He mixes 12·58 litres of water with 4·6 litres of one weed killer and 2·15 litres of another weed killer. How much spray does he have altogether?

- Use standard column procedures to add and subtract whole numbers and decimals...
- Solve word problems and investigate in a range of contexts ...
- Identify the necessary information to solve a problem ...

Addition and subtraction word problems 3

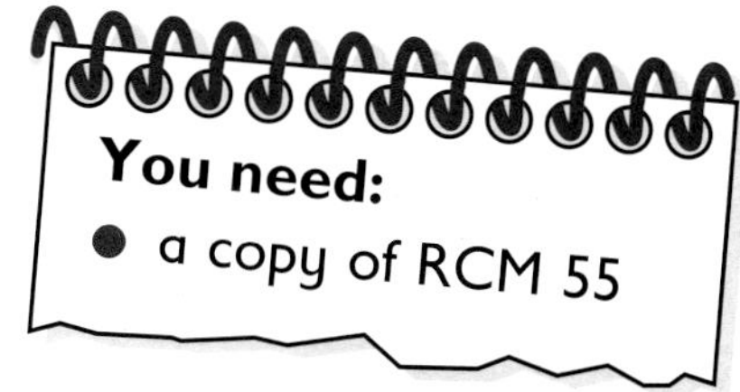

TRIP A

1 On Monday 18th June David travelled from London to Bristol to visit a client and then on to Birmingham where he stayed the night. How far did David travel?

2 On 19th June David pays £67 for his room, £15·18 for dinner the previous evening and £7·26 for his breakfast. He pays with a £50 note and two £20 notes. How much change does he receive?

3 David visits a client in Birmingham in the morning and then drives on to Nottingham to meet another client. That evening he drives back to London. How far does David travel today?

TRIP B

4 On Wednesday 27th June David drives to Bristol to visit his client again and then on to Cardiff to see another client. That evening he drives back home to London. How far does David travel?

TRIP C

5 On Monday 9th July David drives to Leeds. On the way he stops at a service station for lunch. If he has travelled 215·6 km already, how far does he have to travel after lunch before he reaches Leeds?

6 Before David left London he filled up his petrol tank. The tank holds 48 litres. At Nottingham he fills up again with 23·9 litres. How much petrol has he used between London and Nottingham?

7 On Monday 9th July David stays the night at a hotel in Leeds. He spends all Tuesday with a client who he takes out to dinner at his hotel, and stays a second night in the hotel. On Wednesday morning he pays his hotel bill. The bill comes to £216·37. The room charges cost £130 and breakfast £15. The only other item on the bill was for dinner. How much was dinner?

8 On Wednesday he drives from Leeds to Edinburgh. He arrives at lunch time, visits a client and drives onto Glasgow, where he spends the night. How far has David travelled today?

9 Since David left London on Monday 9th July how many kilometres has he driven?

10 On Wednesday he attends a computer fair. Entrance costs £28. At the fair he buys a new computer costing £2463·73 and a new programme costing £438·29. As he has bought both the computer and programme he gets a discount of £44·83. How much did David spend altogether?

11 David stays Wednesday night in Glasgow and on Thursday morning leaves for London. He goes via Liverpool and arrives in London that evening. How far has David travelled today?

12 The next morning, David fills up his petrol tank. This costs him £25·43. On working out his petrol expenses for the trip he has three other bills, one for £20·42, another for £16·32 and a third for £19·87. How much did David spend altogether on petrol?

Multiplying whole numbers

1 Approximate first, then use a written method to work out the answers to these calculations.

a	643×78	b	486×59	c	758×68		
d	794×45	e	563×82	f	382×97		
g	109×76	h	817×36	i	249×22		
j	918×23	k	439×72	l	163×59		
m	545×48	n	806×34	o	282×58	p	627×96
q	354×87	r	876×66	s	879×64	t	708×87

Example

Approximate: 49000

	637
	× 76
637 × 70	44590
637 × 6	+ 3822
	48412
	1 1

2 Look at the your answers to question 1.
Check to see if your answers are correct by adding together the digits in your answer.

If you have made any errors, go back and take another look at the question.

Example

e.g. 48412

$4 + 8 + 4 + 1 + 2 = 19$

a	b	c	d	e
15	27	19	18	23
F	g	h	i	j
19	22	18	24	9
k	l	m	n	o
18	23	15	17	21
p	q	r	s	t
18	27	27	24	27

Multiplying decimals with one place

1 Marcus hires long marquees for people to shelter under during sporting events. Approximate, and then use a written method, to work out the area of these marquees.

Example

628·7 m × 8 m = ▪

Approximate: 4900

628·7 m × 10   6287 × 8 = 50296

50296 m ÷ 10 = 5029.6 m^2

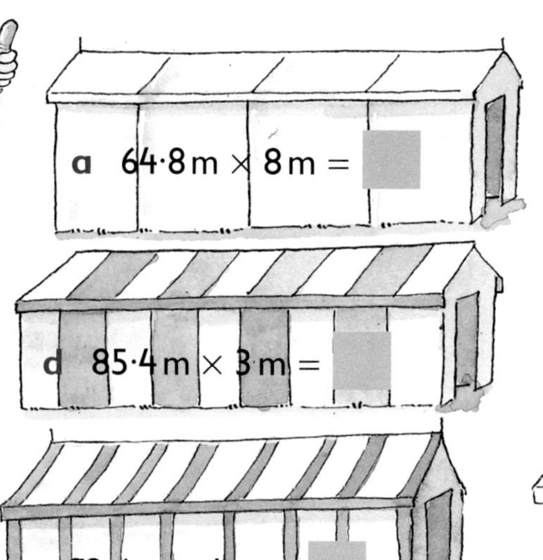

a 64·8 m × 8 m = ▪

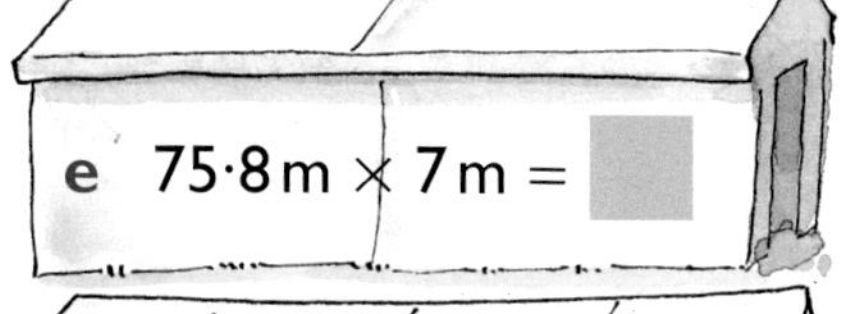

b 112·5 m × 6 m = ▪

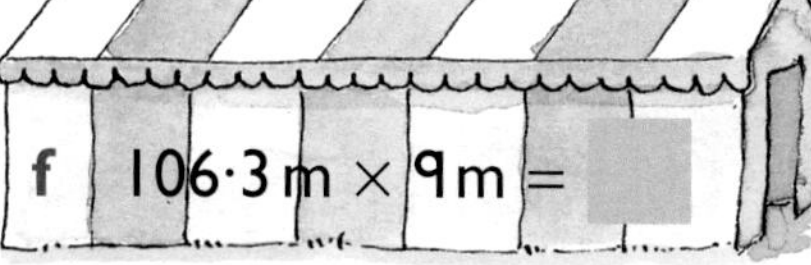

c 97·9 m × 5 m = ▪

d 85·4 m × 3 m = ▪

e 75·8 m × 7 m = ▪

f 106·3 m × 9 m = ▪

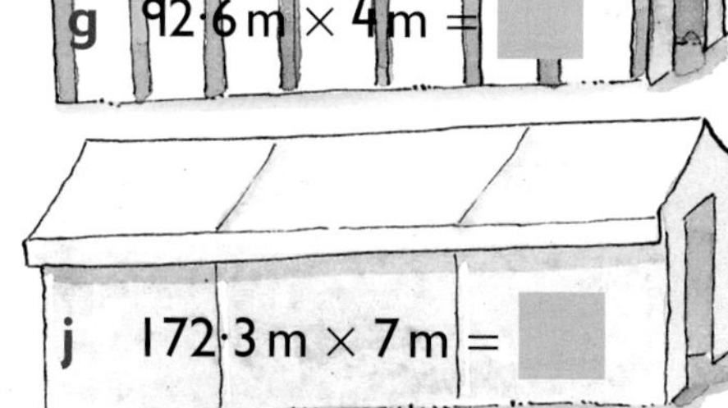

g 92·6 m × 4 m = ▪

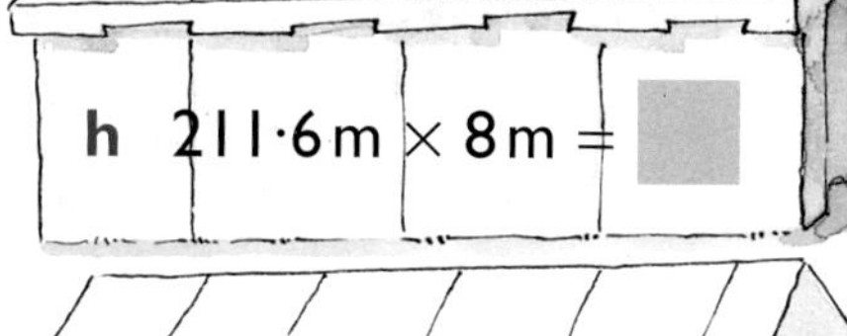

h 211·6 m × 8 m = ▪

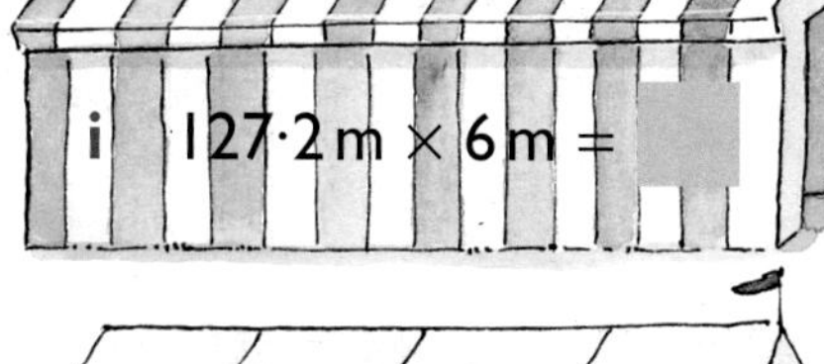

i 127·2 m × 6 m = ▪

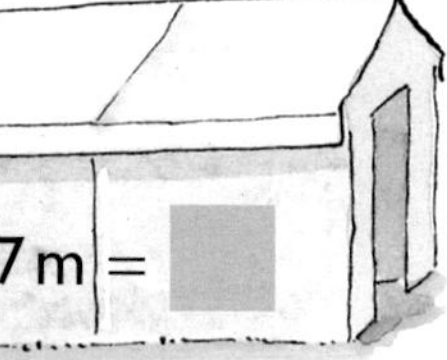

j 172·3 m × 7 m = ▪

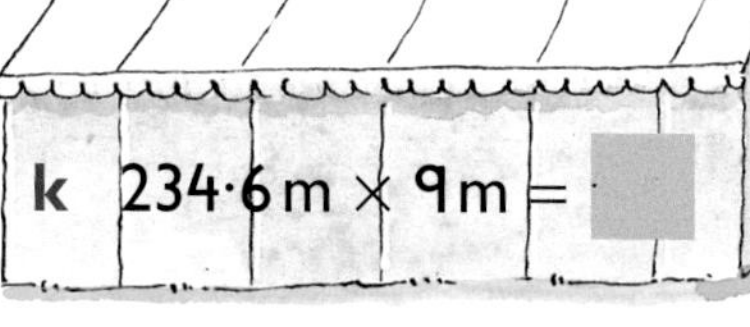

k 234·6 m × 9 m = ▪

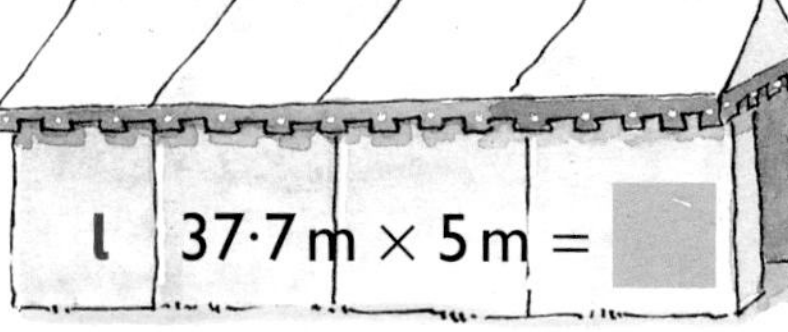

l 37·7 m × 5 m = ▪

2 Make a multiplication.

- Choose a number from the marquee.
- Roll the dice.
- Now multiply the two numbers together.
- Your teacher will tell you how many calculations to make.

You need:

- a 0–9 dice

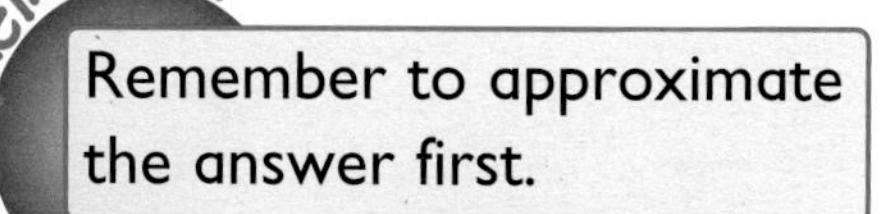

Remember

Remember to approximate the answer first.

Multiplying decimals with two places

Approximate, then use a written method to work out the total air fare cost for the following families.

FLY-THE-WORLD TRAVEL

	Return from		Return from
BRUSSELS	£59·74	NEW YORK	£198·42
PARIS	£62·89	LOS ANGELES	£261·28
AMSTERDAM	£73·67	BEIJING	£386·54
DUBLIN	£85·56	SINGAPORE	£409·79
NICE	£94·35	SYDNEY	£524·35

*Fixed prices, regardless of age

a The Gordan family of 4 is going to Paris.

b The Hearst family of 6 is going to Amsterdam .

c The Vakharia family of 3 is going to Brussels.

d The Molyneaux family of 7 is going to Nice.

e The Schmulian family of 5 is going to Dublin.

f The Jurgensen family of 8 is going Los Angeles.

g The Taylor family of 6 is going to Beijing.

h The Tong family of 7 is going to Singapore.

i The Teuten family of 4 is going to New York.

j The Clarke family of 9 is going to Sydney.

Dividing whole numbers

You need:

- pencil and paper (per child)

Dividing whole numbers

A game for 2–4 players

- Take turns to choose a number from the flag and a number from the sandcastle, for example, 489 and 64.
- The player whose turn it is then divides the 3-digit number by the two-digit number, for example:

Example

```
64)489
  – 320     5 × 64 = 320
    169
  – 128     2 × 64 = 128
     41
 Answer: 7 r 41
```

- The other player or players check to see that the answer is correct.
- The score for that player is the remainder, i.e. 41.
- No two players can choose the same numbers to make a calculation.
- Stop after 12 rounds and add together all your remainders.
- The winner is the player with the largest total.

Variation

- The winner is the player with the smallest total.

44 81
46 55 84
58 26 37 62
97 63 74
29 34 83 95
92 64 17 86 18
72 93 66
51
12 78
31 49 25

- Divide decimals with one place by single-digit whole numbers
- Make and justify estimates and approximations of calculations
- Present and interpret solutions in the context of the original problem…

Lesson 53

Dividing decimals with one place

1 Approximate, then use a written method to work out the answer to these calculations.

a 340·2 ÷ 7 = □   **b** 323·1 ÷ 9 = □

c 464·8 ÷ 4 = □   **d** 216·6 ÷ 6 = □

e 707·4 ÷ 3 = □   **f** 1224·5 ÷ 5 = □

g 765·1 ÷ 7 = □   **h** 2844·9 ÷ 9 = □

i 1266·6 ÷ 6 = □   **j** 869·6 ÷ 8 = □

Example

Approximate: 78

```
8)628·8
 – 560      70 × 8 = 560
   68·8
 –  64      8 × 8 = 64
    4·8
 –  4·8     0·6 × 8 = 4·8
    0
```

Answer: 78·6

2 Make a multiplication.

- Choose a number from the grid.
- Roll the dice.
- Now divide the dice number into the decimal number.
- Your teacher will tell you how many calculations to make.

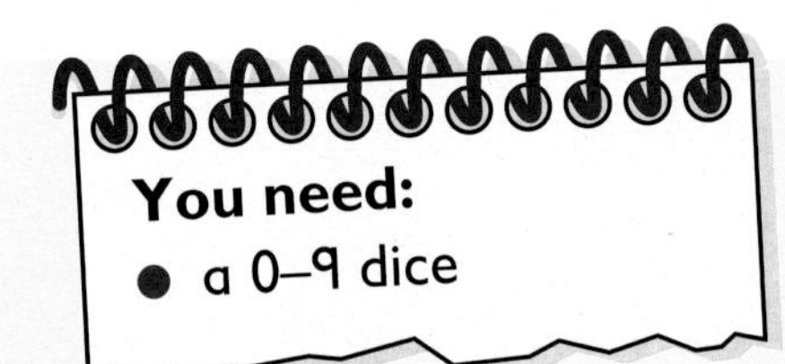

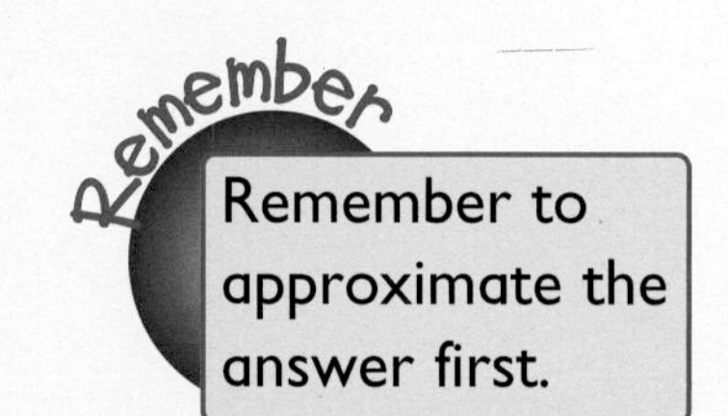

76·7	84·4	261·5	175·6	93·9
542·3	301·8	87·2	268·4	734·1

Puzzle time

Work out the missing digits.

a
```
     8 6 · □
□ ) 6 □ 6 · 2
```

b
```
     3 □ 8 · 4
6 ) □ 2 □ 0 · 4
```

c
```
     □ 4 · 7
4 ) 2 9 □ · □
```

d
```
     1 □ 8 · 3
□ ) 1 □ 3 □ · 7
```

Dividing decimals with two places

Approximate, and then use a written method, to work out how much each market trader makes on average per day.

Example

$$
\begin{array}{rrl}
\text{£}559{\cdot}56 \div 6 = 6\overline{)559{\cdot}56} & & \\
-\ 540 & & 90 \times 6 = 540 \\
19{\cdot}56 & & \\
-\ 18 & & 3 \times 6 = 18 \\
1{\cdot}56 & & \\
-\ 1{\cdot}2 & & 0{\cdot}2 \times 6 = 1{\cdot}2 \\
0{\cdot}36 & & \\
-\ 0{\cdot}36 & & 0{\cdot}06 \times 6 = 0{\cdot}36 \\
0 & &
\end{array}
$$

Answer: £93·26

Multiplication and division word problems 1

1 To improve the look of their city, the local councillors have decided to paint 26 of the city's most important public buildings. They have calculated that they will need on average, 138 litres of paint for each building. How many litres of paint do they need altogether?

2 On the day that the winner is announced the city plans to hold an open-air concert. They have hired 684 chairs and plan to arrange them so that each row contains 18 chairs. How many rows will there be?

3 Shoalhaven won the Beautiful City competition. To commemorate, the Mayor has decided to give each school child a badge. If there are 73 schools in the city and each school needs 438 badges, how many badges will need to be made (rounded to the nearest thousand)?

4 The city councillors have worked out that by winning the Beautiful City competition, 950 new visitors will come to the city each week. How many new visitors is that a year?

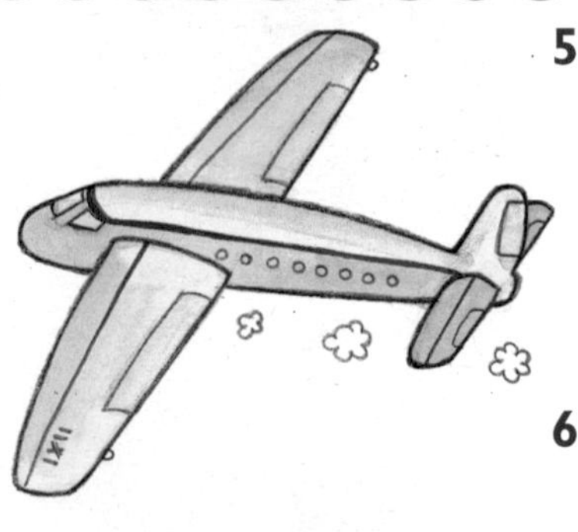

Welcome to Heathrow Airport

5 Between 7:00 a.m. and 7:30 a.m. on 7th August, the baggage handlers at Terminal 3 moved 952 bags. If there were 14 handlers, how many bags did each handler move on average?

6 Flight QF2 flies from London to Sydney. If there are 269 passengers on board and each passenger is allowed up to 32 kg of luggage, what is the maximum weight of luggage they can take on the plane?

7 There are 19 Economy Class check-in desks. In two hours 627 passengers checked in their luggage and were allocated their seats. On average, how many passengers did each check-in desk handle?

8 In six months, Matt the cabin crew attendant made 93 return trips from London to Nice. If each return trip was 343 km, how many kilometres did he travel?

Cinema World

9 Cinema World has 38 cinemas around the country. On average, 437 people visit each cinema each day. What is the total number of people who visit Cinema World cinemas each day?

10 Saturday is the busiest day of the week with an average of 792 people visiting each cinema. The average person spends £14 in the cinema on their ticket, food and drink. Approximately how much money does each cinema take on Saturday?

11 Mondays are the quietest days. Only 4636 people in total visit the 38 Cinema World cinemas on that day. How many people visit each cinema on average on Monday?

12 One of the cinemas is not making much money. It is opened every day of the week but it takes only £5432 each week, and each person visiting it only spends £8 on average. How many people visit this cinema each day, on average?

- Multiply and divide decimals with one or two places by single-digit whole numbers
- Solve word problems and investigate in a range of contexts …
- Identify the necessary information to solve a problem …

Multiplication and division word problems 2

1 In 1997, Andy Green broke the sound barrier in his car Thrust SCC, powered by 2 Rolls Royce engines, setting a land speed record of 1149·3 km per hour. How far would he travel in 10 minutes at this speed?

2 The average person in England uses 672·43 litres of water each day. How much does a family of 5 use?

3 The top speed of a giant tortoise is 0·27 km per hour. In metres, how far can it travel in 7·5 minutes?

4 Krishna earns £375·00 each week. After tax has been deducted, she takes home £298·55. How much does she have to live on each day?

5 Each tyre on a Formula 1 racing car costs £2349·65 and a car, on average, uses 8 tyres during a race. How much is spent on tyres for each car each race?

6 Last week, Johnsons Garage sold 6 cars for £12 962·85 each, including tax and all other costs. How much money did they make last week?

7 The Sydney Opera House is preparing for a concert. It wants to hang 6 large banners from the ceiling. Each banner needs 16·38 m of material. How much material is needed for all 6 banners?

8 The 32 teams competing in the World Cup have been divided into 8 groups. Each group consists of 4 countries. The organisers of Group A have bought 509·6 kg of oranges for the players in all 4 teams to eat during training. What weight of oranges will each team receive?

9 The Sea Princess sails from Florida to Southampton. The journey takes 7 days and uses 27 654·2 litres of fuel. How much fuel does it use each day?

10 Karen is buying new carpet for her living room. The carpet costs £9 per square metre. If the room is 6 m wide by 8·5 m long, how much does the carpet cost for the living room?

11 In 1987 the solar powered car Sunraycer travelled the 2982·42 km from Darwin to Adelaide, Australia, in 6 days. If it travelled the same distance each day, how far did it travel each day?

12 What is the cost of 4 adults, 3 children under 8 and 5 children over 8 to visit the wildlife park?

Maroo Wildlife Park

Adults:	£12·67
Children over 8 years:	£9·32
Children under 8 years:	£6·49

- Multiply and divide three-digit by two-digit whole numbers …
- Solve word problems and investigate in a range of contexts …
- Identify the necessary information to solve a problem …

Multiplication and division word problems 3

This week's best sellers *Book*	*Cost (per book)*	*Number of pages (per book)*
INTO THE WATER by Jonathan Boat	£9·50	104
THREE O'CLOCK by Lucian Watch	£12·65	288
UP A TREE by David Bark	£3·94	72
THE BUTCHER'S REVENGE by Sebastian Pork	£24·10	486
THE SILENT TELEPHONE by Rebecca Ring	£38·29	542

1 Books-R-Us have just ordered 9 copies of *The Silent Telephone*. What is the total cost of the books?

2 *Three O'clock* contains 18 chapters. If each chapter has roughly the same number of pages, how many pages are there in each chapter?

3 Keft Printers in Liverpool printed *Into the Water*. How many pages did they need to print 78 copies of the book?

4 Martin's Books has spent £101·20. They have bought 8 copies of the same book. Which book did they buy?

5 How many copies of *Three O'clock* can David's Desk buy for £90?

6 The Kennington Bookshop ordered 6 copies of *The Butcher's Revenge* and 8 copies of *Three O'clock*. What is the total cost of the books?

7 Gita works 37 hours a week assembling bicycles. She earns £296 a week. How much does she earn an hour?

8 Marcelle works in the canteen. She works 6 hours a day and earns £5·93 an hour. How much does she earn in a 5-day week?

9 Last year, Sports Equipment Limited used 637 tonnes of wood. How much did it use each week if the factory was opened for 52 weeks?

10 Terry works on the production line packaging tennis balls. One box holds 16 tubes of balls and each tube contains 4 balls. If he packs 340 boxes in a day how many balls does he pack?

11 Bob makes cricket bats. He earns £75·60 a day. If he works 8 hours a day, how much does he earn an hour?

12 Sports Equipment Limited produce 768 hockey sticks each week. How many do they make in a 52-week year?

- Use standard column procedures to add and subtract whole numbers and decimals …
- Multiply and divide three-digit by two-digit whole numbers …
- Present and interpret solutions in the context of the original problem …

Puzzles

1 Use the digits 1, 3, 5, 7 and 9 once only to the left of the equals sign to give an answer that has the same digit repeated four times.

▲ ● × ■ ◆ – ♣ = ♥♥♥♥

2 Place the following numbers into groups of three numbers so that the product of all three groups is the same.

5 30 8 6 28 7 3 35 4

3 I think of a number, add 6·2, then multiply by 4.
The answer is 43·2. What is the number?

4 Use these digits to make each of the following statements true.

2 4 8

a 5 ☐ ☐ + ☐ ☐ = 626

b ☐ ☐ 5 – ☐ ☐ = 157

c ☐ · ☐ 5 + 5· ☐ ☐ = 10·29

d 5 ☐ · ☐ + ☐ 5· ☐ = 138

e 5. ☐ ☐ – ☐ ·5 ☐ = 1·24

f ☐ ☐ · 5 – 5 ☐ · ☐ = 32·1

5 Using only the digits 1 and 2 and one of the four operations make the number 610·5.

6 **345 678 123 456**

These numbers are made up of 3 consecutive digits. Choose one of the numbers and reverse its digits.

Find the difference between the two numbers. Repeat for the other numbers above. What do you notice?

Does the same thing work for numbers made up of four consecutive digits?

7 23 × 96 = 32 × 69 = 2208

These 2 two-digit numbers have products that are the same when the digits are reversed.

Can you make another pair?

8 What operation is represented by each ✱

a 411 ✱ 12 ✱ 6 = 205.5

b 411 ✱ 12 ✱ 6 = 822

c 411 ✱ 12 ✱ 6 = 40·25

d 411 ✱ 12 ✱ 6 = 413

9 Using each digit only once per question make each of the following statements true.

1 3 4 6 8

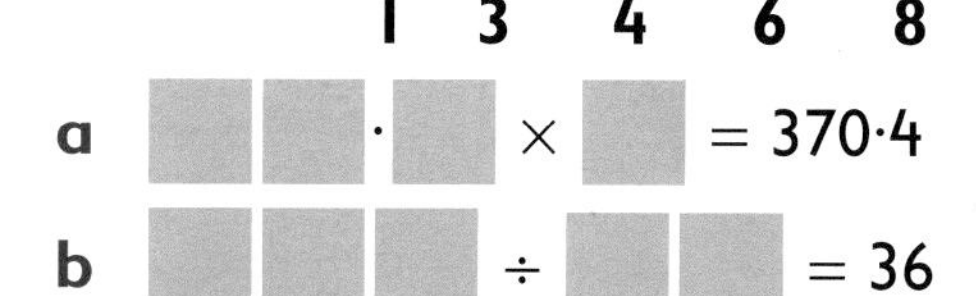

a ☐ ☐ · ☐ × ☐ = 370·4

b ☐ ☐ ☐ ÷ ☐ ☐ = 36

c ☐ ☐ ☐ × ☐ ☐ = 5712

d ☐ ☐ · ☐ ÷ ☐ = 16·2

e ☐ ☐ · ☐ ☐ × ☐ = 144·72

f ☐ ☐ · ☐ ☐ ÷ ☐ = 5·17

10 Find two consecutive numbers with a product of 1406.

Using a calculator 1

Use your knowledge of the order of operations to answer each of these questions. Use a calculator to help you.

You need:
- a calculator

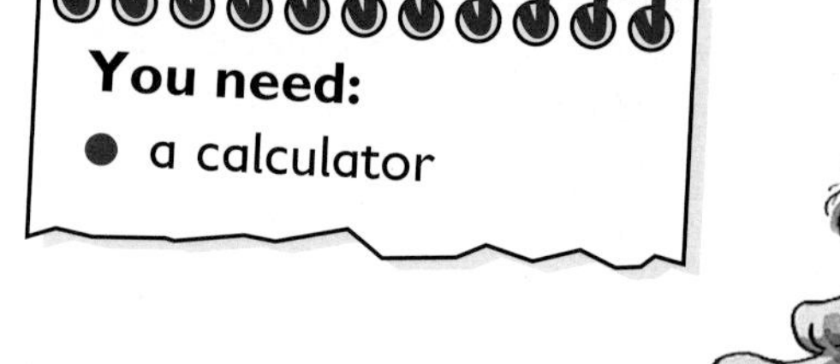

1 a $43 \times 16 + 224 =$ ☐
b $31 - 34 \div 8 =$ ☐
c $132 - 13 \times 7 =$ ☐
d $134 \div 5 + 87 =$ ☐
e $137 \times 7 - 683 =$ ☐
f $824 \div 32 + 369 =$ ☐
g $226 - 897 \div 26 =$ ☐
h $55 + 135 \div 6 =$ ☐
i $947 - 43 \times 17 =$ ☐
j $73 \times 34 + 29 =$ ☐
k $79 \times 84 - 1278 =$ ☐
l $89 + 12 \times 137 =$ ☐
m $903 \div 35 + 169 =$ ☐
n $65 \times 47 - 2723 =$ ☐
o $762 + 43 \times 15 =$ ☐
p $745 - 2295 \div 34 =$ ☐
q $9585 \div 142 + 187 =$ ☐
r $4787 - 29 \times 83 =$ ☐
s $1615 \div 68 - 367 =$ ☐
t $79 \times 163 - 8432 =$ ☐

2 a $43 \times 15 \div 25 =$ ☐
b $23 \times 24 - 18 \times 27 =$ ☐
c $231 \div 6 + 466 \div 8 =$ ☐
d $(69 \times 73) \div (607 - 595) =$ ☐
e $214 \div 8 + 32 \times 17 =$ ☐
f $1170 \div 26 + 113{\cdot}4 \div 9 =$ ☐
g $(351 \div 3) + (142 - 85) =$ ☐
h $26 \times 18 - 24 \times 16 =$ ☐
i $243 + (164 + 278) \div 25 =$ ☐
j $1000 - 7 \times (18 \times 7) =$ ☐
k $45 + 16 \times (137 - 112) =$ ☐
l $(391 \div 17)^2 + 157 =$ ☐
m $(37 \times 25) + (68 \div 4) =$ ☐
n $(28 \times 35) - (186 \div 5) =$ ☐
o $(26 + 86) + (35 \times 8) =$ ☐
p $(5 \times 128) \div (47 - 39) =$ ☐
q $(846 \div 3) + (114 \div 25) =$ ☐
r $614 - 14 \times 31 + 17 =$ ☐
s $85 + 15 \times 37 - 40 =$ ☐
t $(23 \times 14) + (12 \times 17) =$ ☐

Using a calculator 2

You need:
- a calculator
- pencil and paper

1 Use your knowledge of the order of operations to answer each of these questions. Use a calculator to help and make jottings if you need to.

a $43.4 \times 15 + 6 =$ ☐
b $(32.64 - 13.12) \times 69 =$ ☐
c $28.29 \div 4.6 + 83.9 =$ ☐
d $47.1 \times 3.6 - 87.43 =$ ☐
e $43.6 \times 31.9 \div 4 =$ ☐
f $7.43 \times 32 + 16.15 =$ ☐
g $54.9 - 135.27 \div 5.4 =$ ☐
h $(24.76 + 32.69) \times 4.8 =$ ☐
i $137.63 - 397.42 + 341.75 =$ ☐
j $54.82 \times 17.5 + 65.92 =$ ☐
k $87.59 \div 4.75 - 12.84 =$ ☐
l $75.24 \div 45.6 + 13.84 =$ ☐
m $(47.3 + 35.44) \div 8.4 =$ ☐
n $762.45 - 43.75 \times 15.6 =$ ☐
o $(49.59 + 126.9) \div 3 =$ ☐
p $83.4 \times 3.55 - 279.6 =$ ☐
q $321.96 + 465.85 \div 5.5 =$ ☐
r $35.7 \times 49.8 + 362.7 =$ ☐
s $79.4 + 168.92 \times 3.5 =$ ☐
t $56.24 \times 6.5 - 232.46 =$ ☐

2 Use your knowledge of the order of operations to answer each of these questions. Use a calculator to help and make jottings if you need to. Remember to convert measures to the same unit.

a 80% × £4·65 + £9·80 = £ ☐
b 345 g × 112 – 24·3 kg = ☐ kg
c 361·9 cm ÷ 8·8 + 38·75 mm = ☐ cm
d £13·42 + 73 × 49p = £ ☐
e 4 min 43 s + 6 × 1 min 23 s = ☐ min ☐ s
f 654 g + 74·82 kg ÷ 645 = ☐ kg
g (43·63 ml – 5·75 ml) × 7500 = ☐ l
h 105 m – 7490 × 14 mm = ☐ m
i £142·40 ÷ 160 + £1·78 = £ ☐
j 1 day 4 h 53 min 45 s ÷ 365 = ☐ min ☐ s
k 753·6 l ÷ 4800 + 38·5 cl = ☐ cl
l (4·36 kg + 270 g) × 154 = ☐ kg
m 63·5 cl + 42% × 5·75 l = ☐ l
n 43 s × 7345 = ☐ h ☐ min ☐ s
o 875 g × 98% – 0·45 kg = ☐ g
p 4·42 l ÷ 1·25 – 270·4 cl = ☐ ml
q £190·86 – 29p × 581 – £18·65 = £ ☐
r 13·63 l + 46·3 cl – 34 ml × 392 = ☐ ml
s 692 × (27·8 g + 17·2 g) – 25·86 kg = ☐ kg
t 40 m – (12·7 cm + 98 mm + 38 mm) × 150 = ☐ m

Using a calculator to solve problems

You need:
- a copy of RCM 57
- a calculator

Use the information from the FOR SALE section of RCM 57: Wonderhome Estate Agents to answer these questions.

1 The price of the studio has just been increased by 11%. What is the new price of the studio?

2 The Thompson family want to buy the 3 bedroom house. The most they can afford to pay is £155 600. How much more money does the Thompson family need to raise?

3 There has been a lot of interest in the 3 bedroom flat. As a result, the owner has decided to increase the price by 7%. What is the new price of the flat?

4 The owner of the 2 bedroom flat needs to sell the flat quickly so she has decided to drop the price by 4%. What is the new price of the flat?

5 What is the difference in price between the 3 bedroom house and the 3 bedroom flat?

6 The people who have bought the 6 bedroom house could not afford the asking price. So they have sold the tennis court and part of the garden to their new neighbours. If they sold the tennis court for £18 700 and part of the garden for £32 500. How much did they have to pay for the rest of the property?

Use the information from the FOR RENT section of RCM 57: Wonderhome Estate Agents to answer these questions.

7 What is the cost of renting the 4 bedroom flat for a year?

8 If there are on average 4 weeks a month, how much does the 3 bedroom flat cost per week?

9 Rached earns £2050 a month. From this he spends about £120 a week on food, transport, entertainment and other expenses. If there are on average 4 weeks per month, can Rached afford to rent the 2 bedroom house? If so, how much would he have left over per month? If not, how short would he be per month?

10 How much does it cost to rent the 4 bedroom house for a year?

11 Two sisters want to rent the 2 bedroom flat. Each month Melissa earns £1250 and spends about £80 a week on food, transport and other expenses. Bridget earns £1115 a month and spends on average £70 a week.

a If there are on average 4 weeks per month can the sisters afford to rent the flat? If so, how much would they have left over per month? If not, how short would they be per month?

b If each sister pays half the rent each month. How much money does each sister have left at the end of the month?

12 Two students want to rent the 1 bedroom flat for a year. If each student pays half the rent how much will they each have to pay over the year?

Checking your answers

Work out the answers to these calculations. Then use a different method to check your answer.

1

$9528 + 3742 = \square$

Now check your answer

2

$829 \div 37 = \square$

Now check your answer

3

$693{\cdot}63 - 58{\cdot}52 = \square$

Now check your answer

4

$582 \times 74 = \square$

Now check your answer

5

$192{\cdot}3 \div 3 = \square$

Now check your answer

6

$62{\cdot}34 \times 6 = \square$

Now check your answer

Letters for numbers

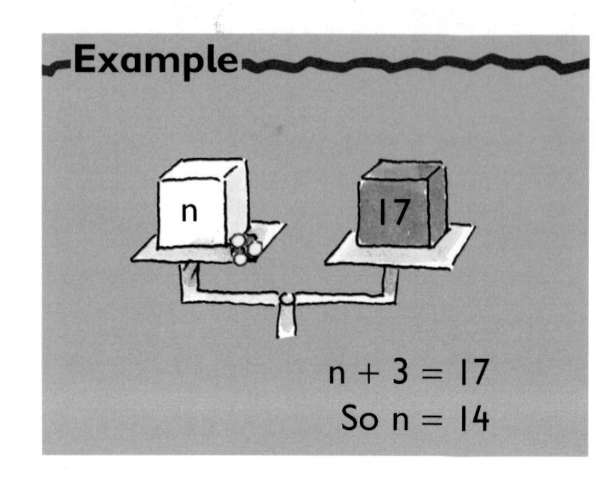

1 Write the number each letter stands for to make each machine balance.

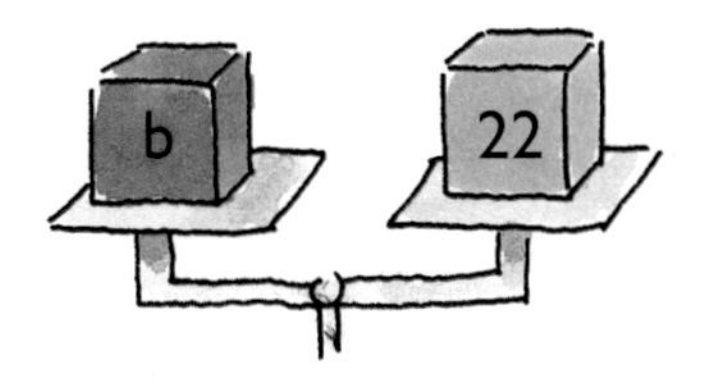

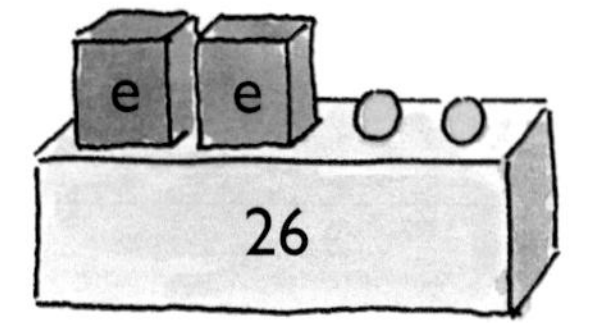

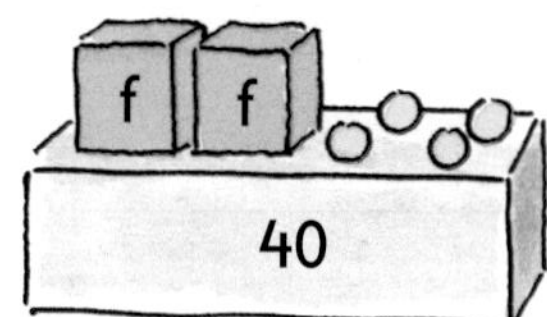

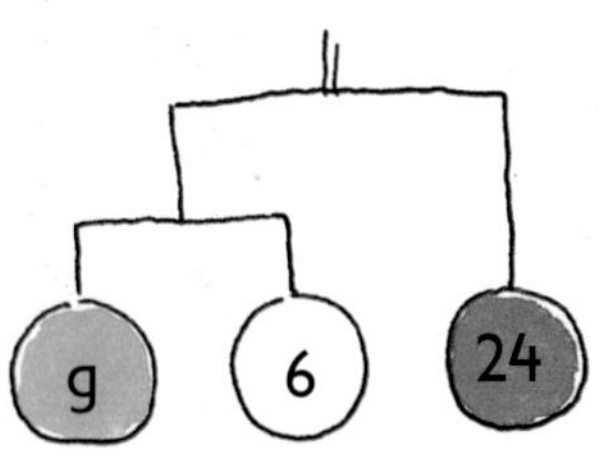

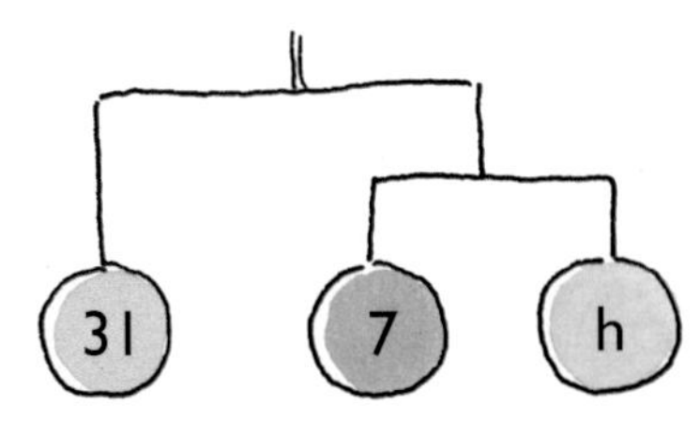

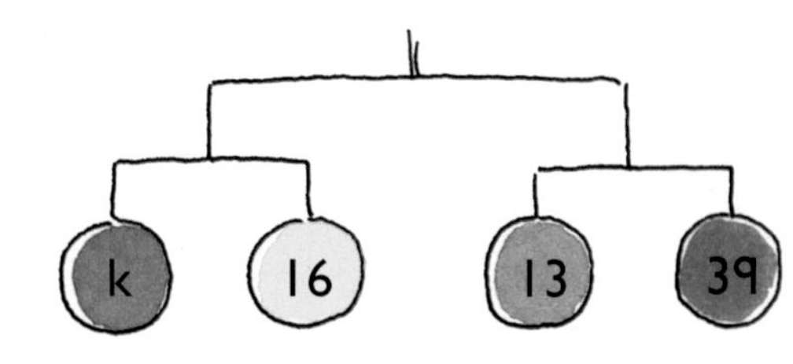

2 For each bag of marbles, write an equation and find the number that the letter stands for.

Take 7 out to leave 16.

Remove 15 to leave 42.

Add 9 more to make the total 42.

Subtract 2 lots of 6 to leave 27.

3 A box has 20 shapes. They are squares (s) or triangles (t). Copy and complete the table to show some of the possible amounts.

s	9		6		11		8		3
t		13		18		5		7	

4 The chart shows how many of each shape are in a box. You can write the total of squares and triangles as an equation. $s + t = 79$

s	t	r	p	h
41	38	62	54	26

a Find and write 9 more equations by combining 2 sets of shapes.

b What if you also had 18 kites?

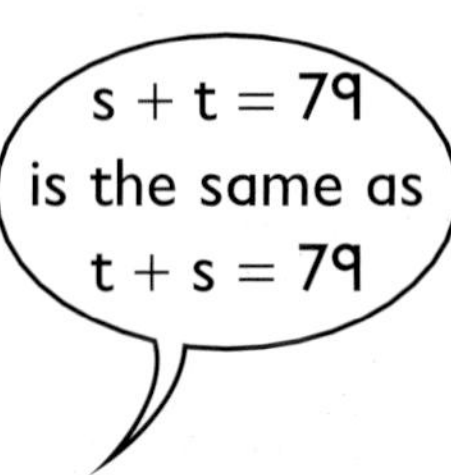

Expressions and equations

Example

s + 14

$s + 14 = 8 + 14$
$= 22$

1 Calculate the value of each expression.

$s = 8$

a $s + 2$ **b** $s - 6$ **c** $s - 5$ **d** $12 + s$

2

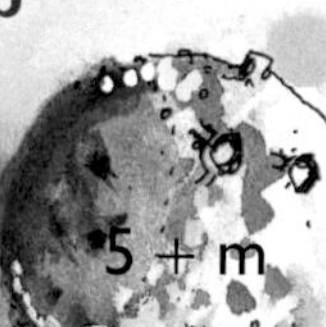
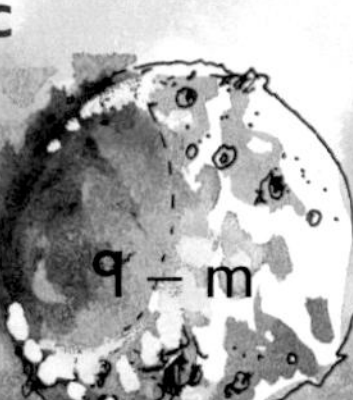
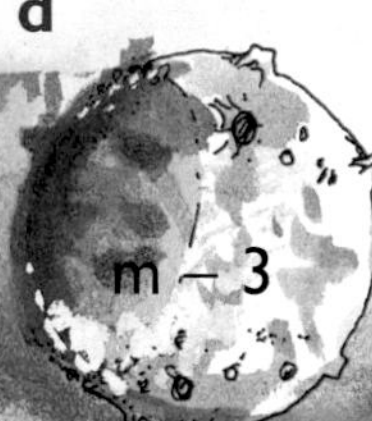
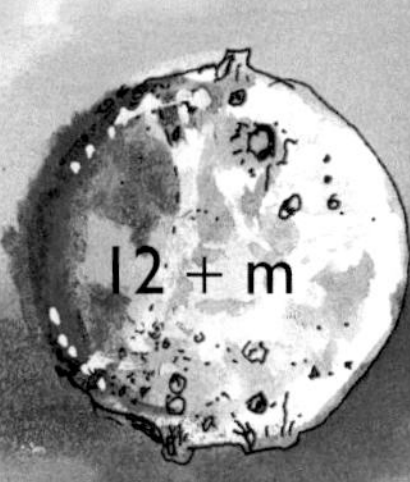

$m = 3$

a $m + 2$ **b** $5 + m$ **c** $9 - m$ **d** $m - 3$ **e** $12 + m$

3

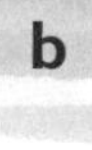

$c = 6$

a $4c$ **b** $7c$ **c** $5c$ **d** $12c$ **e** $9c$

4

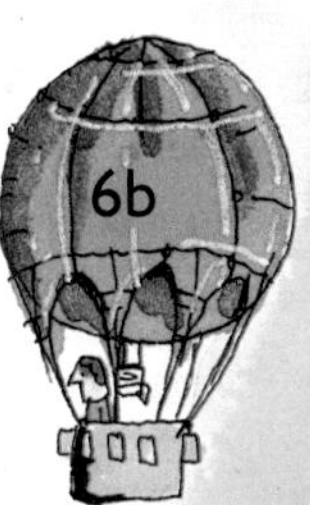
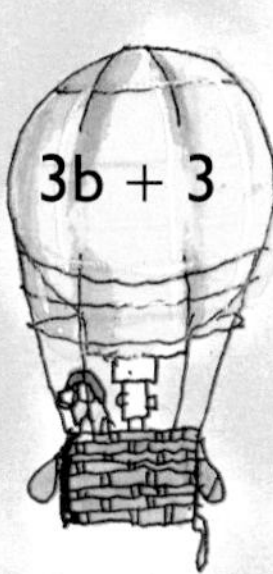
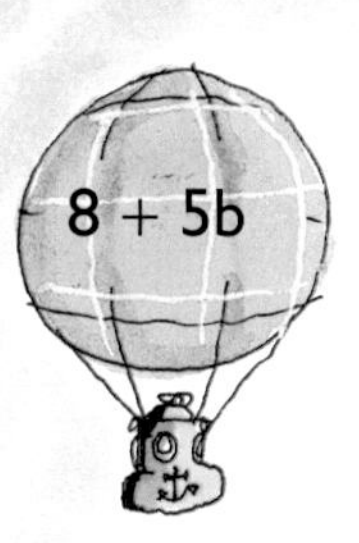
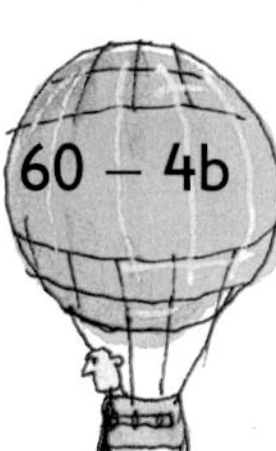

$b = 10$

a $6b$ **b** $3b + 3$ **c** $8 + 5b$ **d** $60 - 4b$ **e** $9b - 12$

5 Calculate the value of each expression when $r = 4$ and $s = 2$

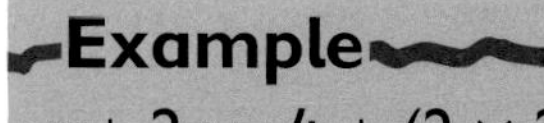

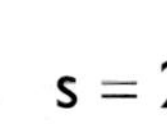
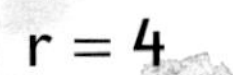

Example

$r + 2s = 4 + (2 \times 2)$
$= 4 \times 4$
$= 8$

$2r - s = (2 \times 4) - 2$
$= 8 - 2$
$= 6$

a $r + 2$ **b** $2r$ **c** $2s$

d s^2 **e** $2r + s$ **f** $2r - s$

g $r + 2s$ **h** $2r - 2s$ **i** $2r + 4s$

j $3r - 2s$ **k** $5r - 5s$ **l** $r - 2 + s$

Collecting like terms

Remember Collect like terms.

Example
$p + p + 4 = 2 \times p + 4$
$= 2p + 4$

1 Write an expression to show what each lorry holds.

a

b

c

d

e

f

2 Match each ticket to the correct bus route.

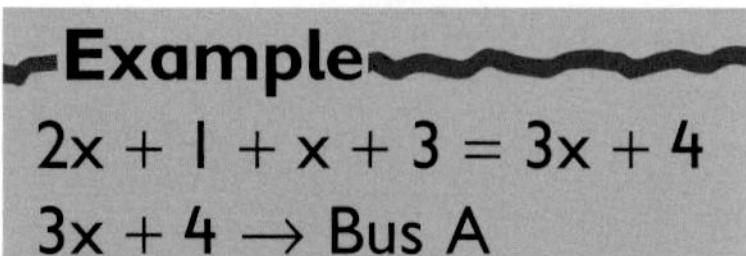
Example
$2x + 1 + x + 3 = 3x + 4$
$3x + 4 \rightarrow$ Bus A

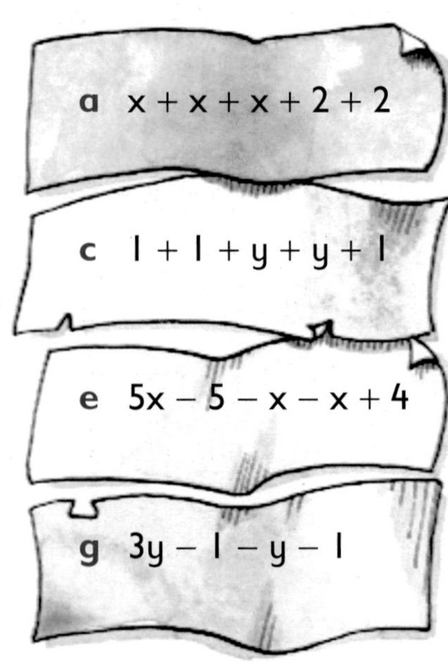

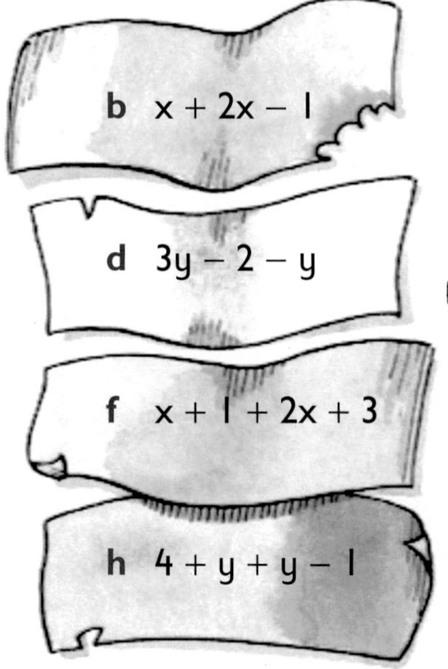

3 Simplify these expressions.

a $2(m + 3)$ b $4(n - 2)$ c $6p \div 2$
d $5(q - 4)$ e $10(r + 7)$ f $6(2s + 3)$
g $3(m + 5)$ h $7(n - 7)$ i $9(p + 2)$
j $10(q + 10)$ k $8(r - 5)$ l $20(s + 3)$

Remember Multiply everything inside the bracket by what is outside.

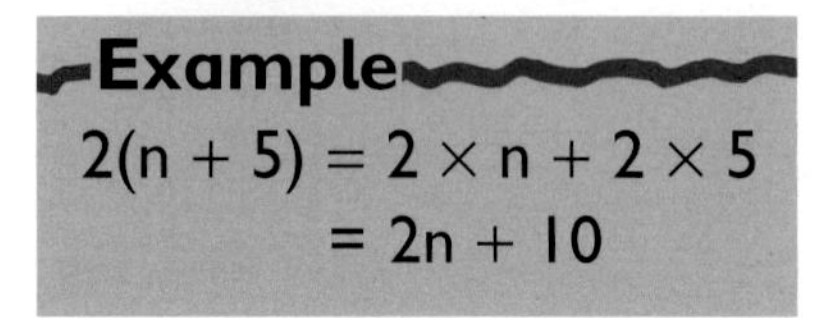
Example
$2(n + 5) = 2 \times n + 2 \times 5$
$= 2n + 10$

Brick walls

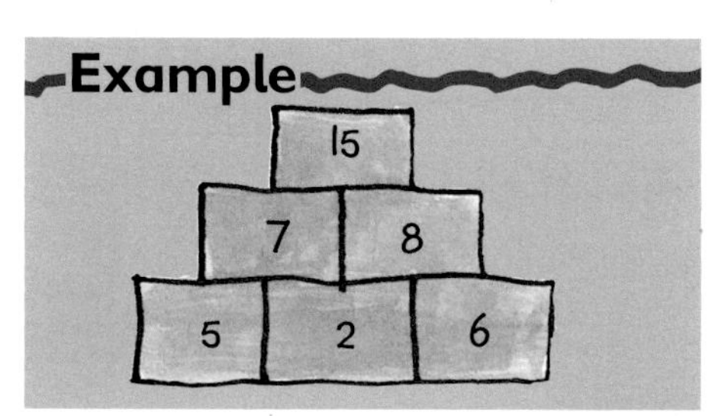

1 The number on each brick is the result of adding the numbers on the 2 bricks beneath it. Find the missing numbers in these brick walls.

a

b

8

5 1

c

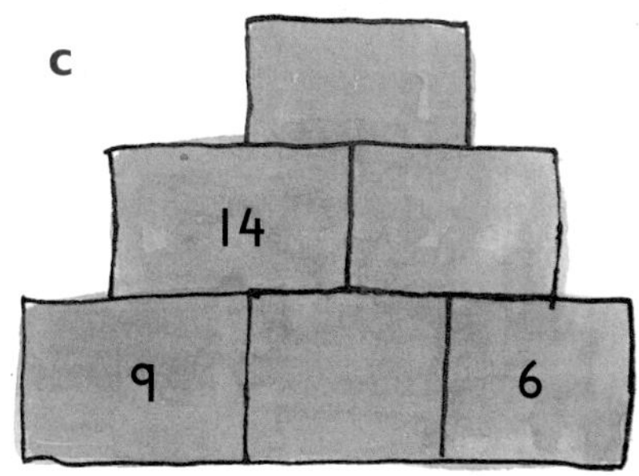

d

e

f

2 In this brick wall the number in the middle brick is the sum of the two lettered bricks on either side of it.

a Copy and complete these relationships.

A + B = 7

C + D = ☐

☐ + ☐ = 12

☐ + ☐ = 4

b Find numbers for A, B, C and D. Copy and complete the table.

A	3				0
B		5			
C			3		
D				9	

Expressions and formulae

Example

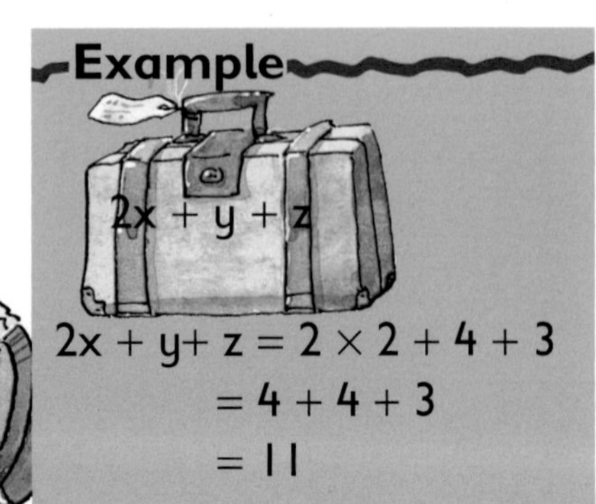

$2x + y + z = 2 \times 2 + 4 + 3$
$= 4 + 4 + 3$
$= 11$

1 Substitute the values for x, y and z to find the weight of each suitcase.

$x = 2 \quad y = 4 \quad z = 3$

a

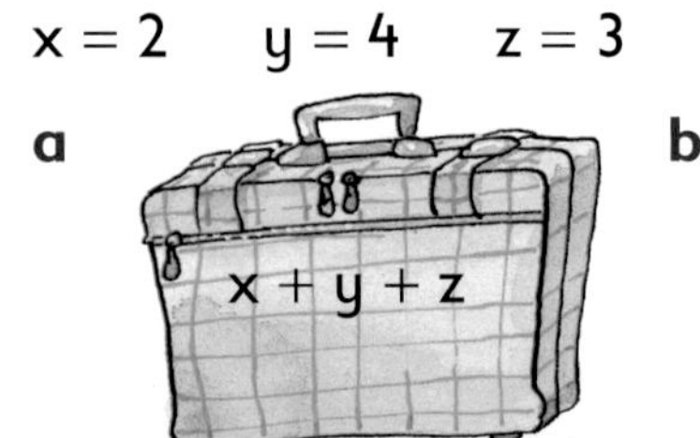

b $9z - x$

c $3x + 2y - z$

d

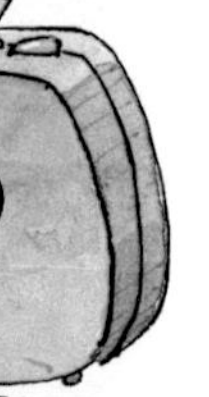

e

f

g

h

i

2 a Find the values for a and b. Copy and complete the spreadsheet.

a + b	a – b	a =	b =
8	2	5	3
7	3		
13	5		
12	4		
11	7		
15	3		

Example

The sum of two numbers is 8. $a + b = 8$
Their difference is 2. $a - b = 2$
Try a = 5 and b = 3.
5 + 3 = 8 ✓ and 5 – 3 = 2 ✓

b If $a \times b = 36$ and $a \div b = 4$, what numbers do a and b stand for?

3 In this street the house numbers increase by 2 each time.
The number of the first house is x.
The sum of the house numbers is 70.

a Write the numbers of all 5 houses in terms of x.

b Find the number of the first and last house,

c What if the house numbers increase by 4 each time?
Find the number of each house.

- Use simple formulae from mathematics and other subjects; substitute positive integers into simple linear expressions and formulae and, in simple cases, derive a formula
- Solve word problems and investigate in a range of contexts: number, algebra

Lesson 68

Deriving formulae

1 *The number of miles = 60 times the number of gallons.*

a Find a formula for the number of miles (m) for each gallon (g) of petrol.

b Use the formula to find how far each rider cycled.

i 4 gallons   ii 6 gallons   iii 1·5 gallons   iv 3·2 gallons   v 0·8 gallons

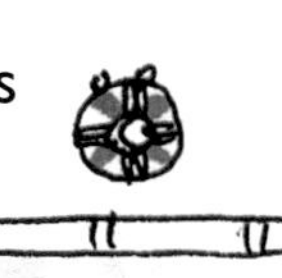

2 The length of a swimming pool is twice its width. Find the perimeter and surface area of a swimming pool when:

a w = 5 m   **b** w = 8 m

c w = 13 m   **d** w = 25 m

3 The slabs in these paths are regular polygons.

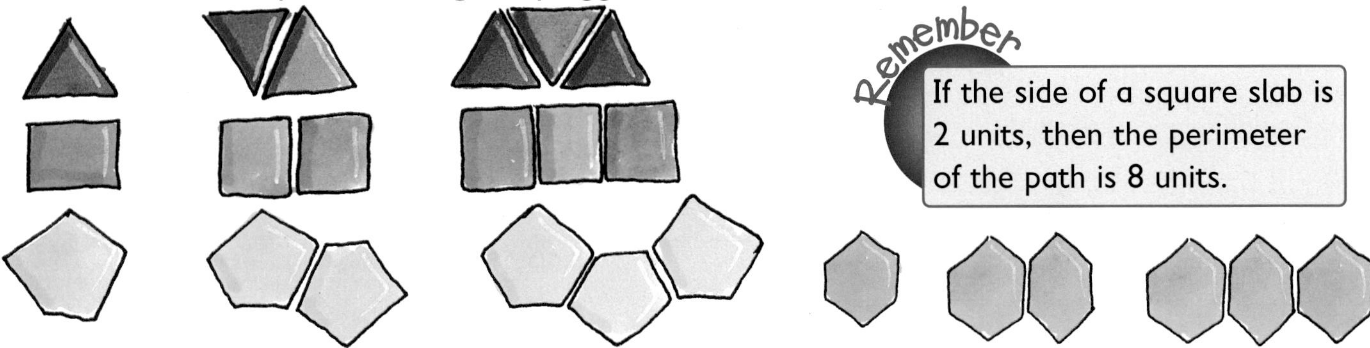

Remember If the side of a square slab is 2 units, then the perimeter of the path is 8 units.

a Copy and complete the table for these polygonal slabs.

	Number of slabs used								
	1	2	3	4	5	6	7	8	n
Perimeter of path in units	3	4							P = n + ▢
	4		8						
	5								
	6								

b Use your results to derive a formula for a path of regular octagon slabs.

c Find the perimeter, in units, of a path of 25 octagon slabs.

Next in line

Example
8, 16, 24, 32, 40, 48, 56, 64, 72
Rule: Each term is a multiple of 8.

1 For each sequence:
- Write the next 5 numbers.
- Explain the rule.

a 6, 12, 18, 24, …

b 550, 500, 450, 400, …

c 0·4, 0·8, 1·2, 1·6, …

d 144, 132, 120, 108, …

e 6, 3, 0, −3, …

f 8·4, 7·7, 7·0, 6·3, …

Example
69, 68, 66, 63, 59, …
Each term is made by subtracting 1 then 2 then 3, …

2 Work out how each sequence is made. Continue the sequence to the 10th term. Explain the rule.

a 1, 3, 6, 10, 15, …

b 1, 2, 4, 7, 11, …

c 2, 4, 8, 16, 32, …

d 60, 59, 57, 54, 50, …

e 100, 98, 94, 88, 80, …

f (5, 8), (10, 16), (15, 24), …

3 The rule for a sequence can have two steps.

Example
Rule: Each term is a multiple of 5 plus 1.

Counting numbers	1	2	3	4	5	6	7	8	9	10
Multiply by 5	5	10	15	20	25	30	35	40	45	50
Add 1	6	11	16	21	26	31	36	41	46	51

New sequence is 6, 11, 16, 21, 26, … 51

Use the counting numbers 1 to 10. For each rule write the first 10 terms.

a A multiple of 2 plus 1

b A multiple of 5 minus 1

c A multiple of 7 plus 2

d A multiple of 10 minus 2

4 Copy and complete the next two shapes in these sequences.

a

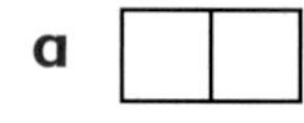

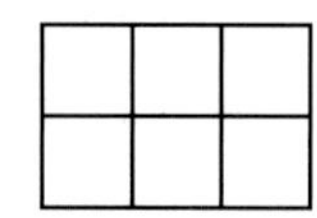

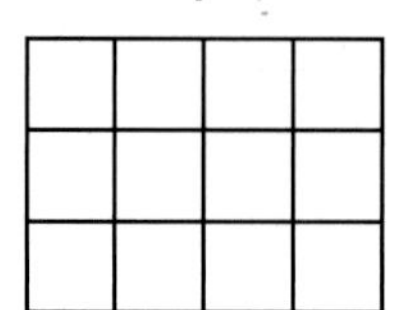

b

c

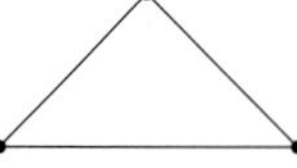

Generating terms

Example
1st term: 10
Rule: add 3
Sequence: 10, 13, 16, 19, 22, 25
Each term is 3 more than the one before.

1 Write the next 5 terms of these sequences then describe each sequence in words.

	1st term	Term-to-term rule
a	6	Add 4
b	100	Subtract 6
c	2	Double
d	3	Multiply by 10

2 Use the rule to write the first six terms of a sequence where:

Term-to-term rule: add 6.

a all the numbers are multiples of 3

b all the numbers are odd

c all the numbers are even but are not multiples of 3.

3 Copy and complete the table.

	Position	1	2	3	4	5	6	n
a	Term	5	6	7	8			
b	Term	7	14	21				
c	Term	3	5	7				
d	Term	10	15	20				
e	Term	8						n + 7
f	Term	4						4n
g	Term	1						2n − 1

4 For each sequence in question 3, use the position-to-term rule to find:

a the 10th term **b** the 100th term

5 The rule for a sequence of numbers is:

Add the two previous numbers.

The first five terms in the sequence are: 1, 2, 3, 5, 8 …
Write the next 4 terms.

Building sequences

You need:
- centicubes
- rods

1 Here are 3 window patterns made with centicubes.

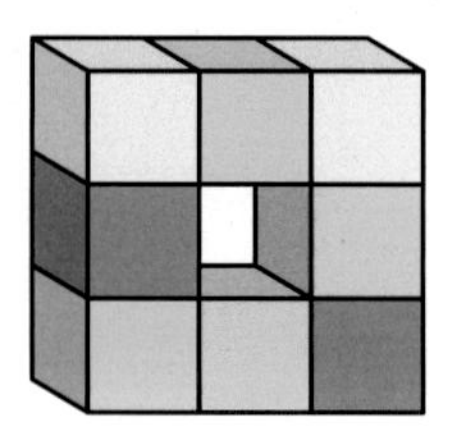
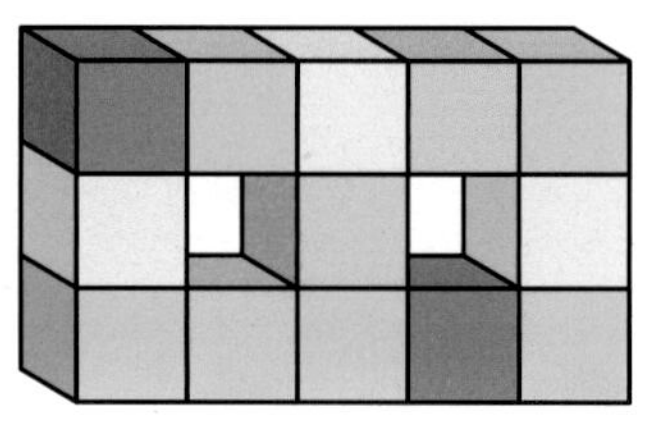
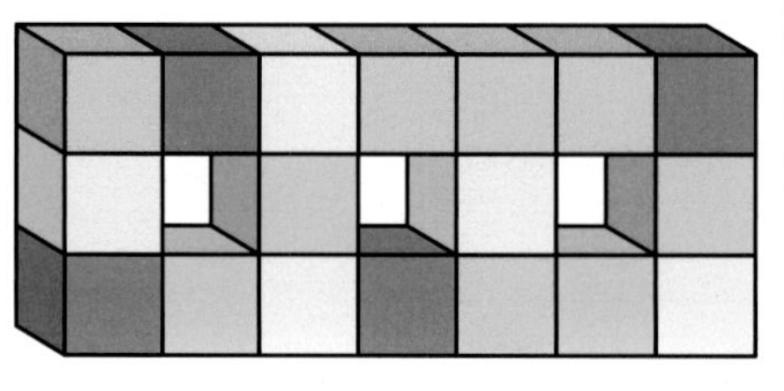

a Make the next 3 patterns in the sequence.

b Copy and complete the table.

Number of windows	1	2	3	4	5	6
Number of bricks	8	13				

c Find a rule for the nth arrangement.

d Justify your rule.

e Use the rule to find the number of bricks in a building with 10 windows.

2 Here are 3 doorway patterns made with rods.

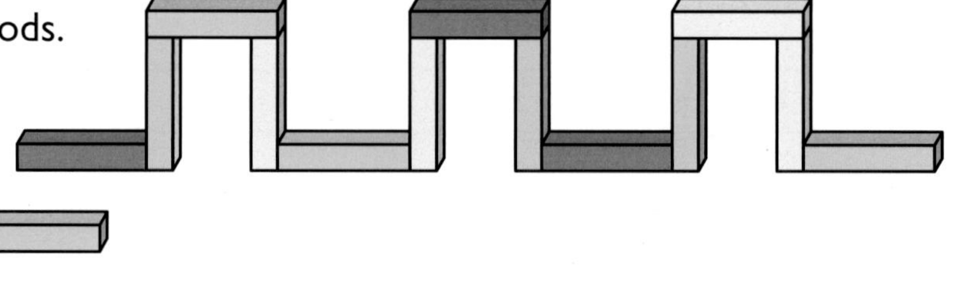

a Make the next 3 patterns in the sequence.

b Copy and complete the table.

Number of doorways	1	2	3	4	5	6
Number of rods	5	9				

c Find a rule for the nth arrangement.

d Justify your rule.

e Use the rule to find the number of rods in a corridor of 10 doorways.

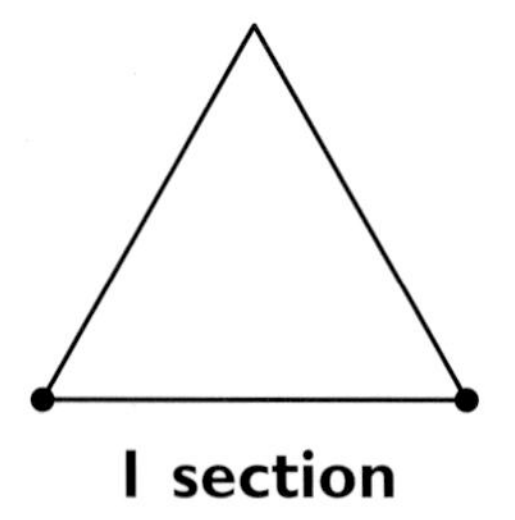
1 section

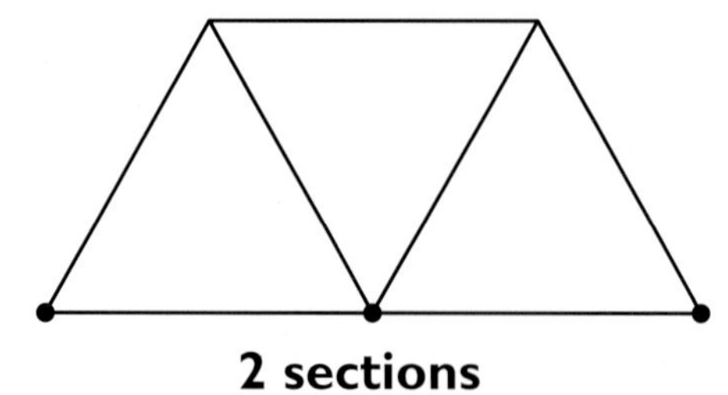
2 sections

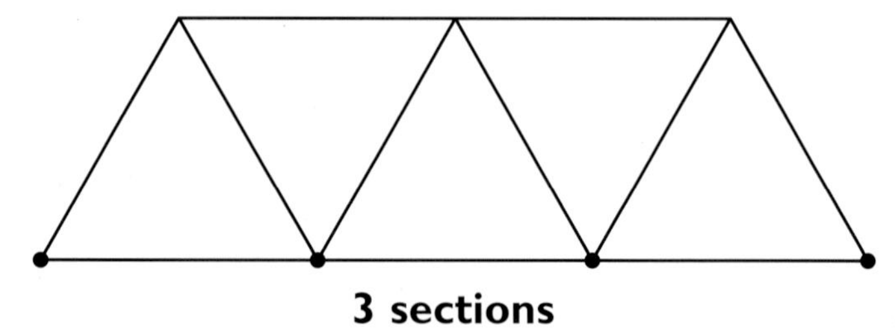
3 sections

3 The sides of a bridge are constructed by joining sections of steel girders.

a Find a rule for the nth section.

b Justify your rule.

Investigating diagonal patterns

This triangle has many patterns.

1st diagonal
2nd diagonal
3rd diagonal
4th diagonal
5th diagonal

1
2 4
3 6 9
4 8 12 16
5 10 15 20 25
6 12 18 24 30 36
7 14 21 28 35 42 49
8 16 24 32 40 48 56 64

1 Copy and complete the table.

Find the rule for the nth term for each diagonal.

Use the rule to find the 20th term in each sequence.

Term	Diagonal				
	1st	2nd	3rd	4th	5th
1	1	4	9	16	25
2	2	6	12	20	30
3	3	8	15	24	35
4					
5					
6					
7					
8					
nth					
20					

2 **a** Describe the sequence of 1st term numbers.

b Write the rule for the nth term.

c Find the 10th term in the sequence of 1st term numbers, then use it to write the first 7 terms of the 10th diagonal.

Inputs and outputs

1 Complete RCM 31, Function machines.

2 weight of beef in kilograms → 40 minutes per kg → plus 20 minutes → roasting time in minutes

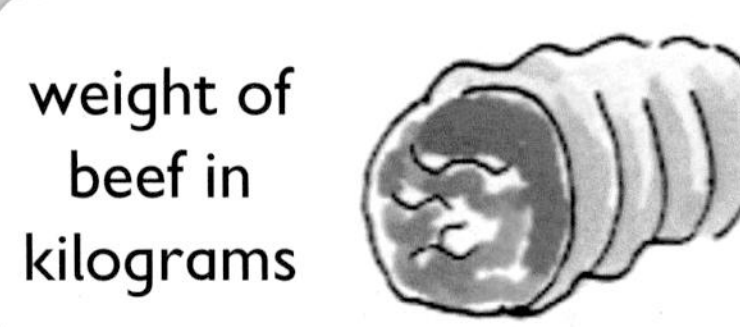

a Copy and complete the table for the roasting time in minutes for joints of beef from 1 kg to 10 kg.

Weight (kg)	1	2	3	4	5	6	7	8	9	10
Roasting time (min)										

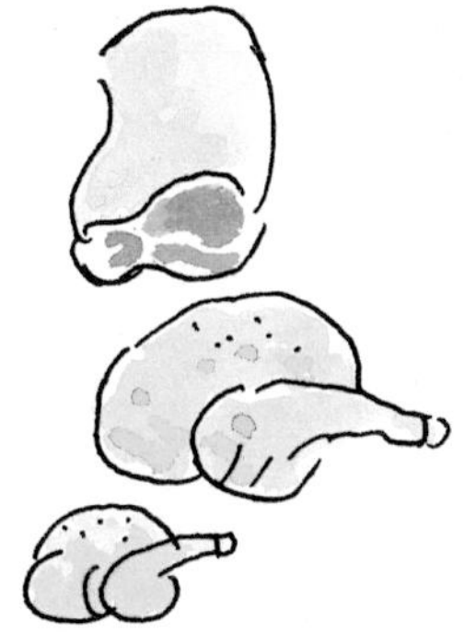

b Work out the roasting time in minutes for these joints of meat from 1 kg to 10 kg.

lamb .... 50 min per kg + 30 min
turkey .... 30 min per kg + 15 min
chicken .... 20 min per kg + 25 min

Weight (kg)	1	2	3	4	5	6	7	8	9	10
Lamb: time (min)										
Turkey: time (min)										
Chicken: time (min)										

3 Find the cost of dinner at the restaurant for 1 to 10 people.

Fontana Restaurant
Dinner Menu
£15 per person
£2 service charge

Hint: Write the rule as a mapping then make a table of inputs and outputs.

4 CPM Taxis use this formula to calculate hire charges.

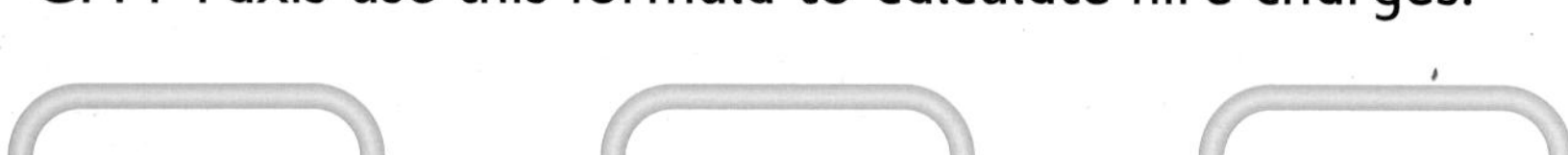

distance → × 90p → + 50p → cost

a Calculate the cost in pounds for taxi hires of these distances:

i 10 miles ii 15 miles iii 25 miles iv 40 miles

b Tim's taxi fare was £11·30. How long was his journey?

Finding the function

Example

4, 5, 6, 7 → ? → 6, 7, 8, 9

Add 2 to the input.

$x \rightarrow x + 2$

1 Write the rule for these machines in words and in symbols.

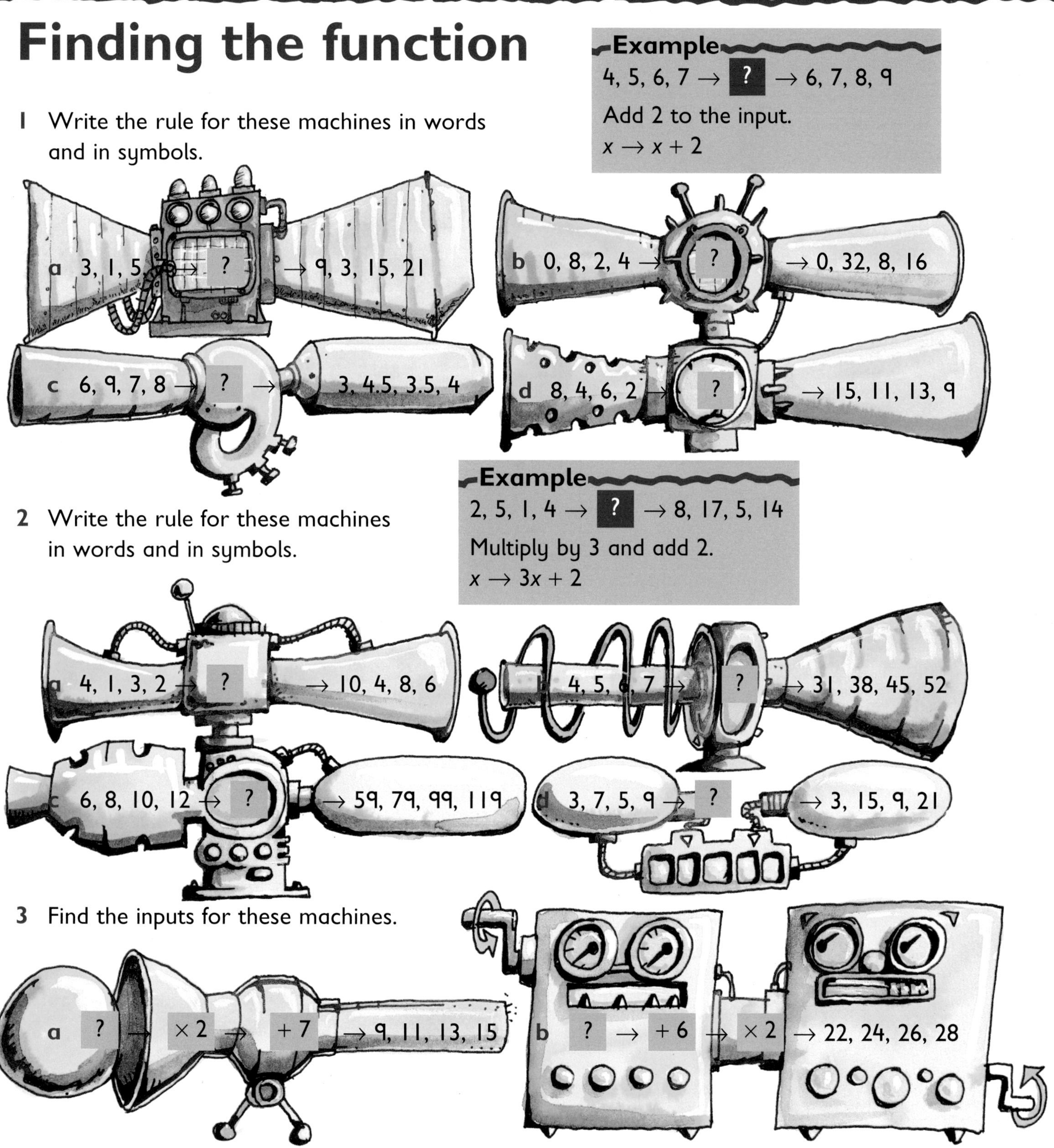

Example

2, 5, 1, 4 → ? → 8, 17, 5, 14

Multiply by 3 and add 2.

$x \rightarrow 3x + 2$

2 Write the rule for these machines in words and in symbols.

3 Find the inputs for these machines.

4 Investigate the chain of numbers you can generate with this rule.
Rule: × → multiply the units digit by 4 → add the tens digit → y

Example

16 → 6 × 4 = 24 → 24 + 1 → 25

Start at 25 and continue the sequence as far as you can go.

Plotting pairs of coordinates

The rule is: $y = x + 1$
If $x = 2$ then $y = 3$
We use the rule to generate pairs of coordinates:
(0, 1), (1, 2), (2, 3), (3, 4), (4, 5) …
We plot the points in a coordinate grid and extend the line.

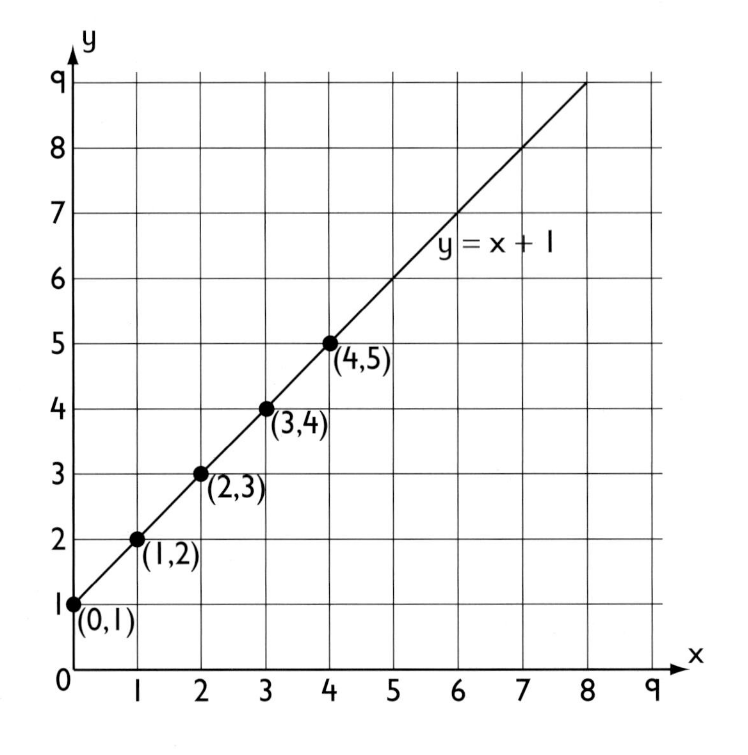

You need:

- a copy of RCM 32
- a ruler
- a sharp pencil

1 For each question, use a different grid on RCM 32.

- Plot the pairs of coordinates.
- Draw a line through the plotted points.
- Extend the line as far as it will go.

a $y = x + 2$
(0, 2), (1, 3), (2, 4), (3, 5), (4, 6) …

b $y = 9 - x$
(0, 9), (1, 8), (2, 7), (3, 6), (4, 5) ...

c $y = x + 4$
(0, 4), (1, 5), (2, 6) …

d $y = 7 - x$
(0, 7), (1, 6), (2, 5) …

2

- Copy and complete the table of values to satisfy the rule $y = x + 1$
- Plot the points on a 4-quadrant coordinate grid.
- Draw a line through the plotted points.
- Extend the line as far as it will go.
- Choose an intermediate point on the line, but not one of those plotted.
  Read off the coordinate pair.
  Check that it fits the rule.
- Repeat for 2 more points.

x	−3	−2	−1	0	1	2	3
y = x + 1	−2	−1					

3 Repeat the steps in question 2 for this table of values.

x	−3	−2	−1	0	1	2	3
y = x + 3				3			

Plotting straight-line graphs

You need:
- a copy of RCM 33
- a ruler
- a sharp pencil

1 a Copy and complete these tables of values.

$y = x$

x	1	2	3	4	5	6
y	1	2				

$y = 2x$

x	1	2	3	4	5	6
y	2	4				

$y = 3x$

x	1	2	3	4	5	6
y	3					

$y = 4x$

x	1	2	3	4	5	6
y	4					

b Use grid 1 of RCM 33 to plot all 4 graphs.

c Draw a line through the plotted points and extend the line in both directions.

d Write the coordinates of the point through which all four lines pass.

e Construct a table of values for $y = \frac{1}{2}x$ and plot the graph on the same grid.

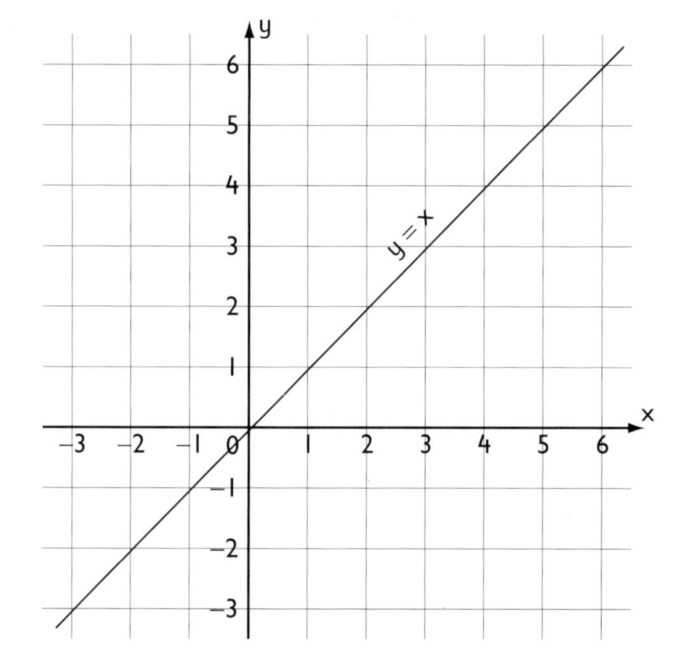

2 On grid 2, the graph of $y = 3$ is drawn.

$y = 3$

x	1	2	3	4
y	3	3	3	3

a Construct tables and on grid 2 plot graphs for: $y = 1$, $y = 5$, $y = 7$ and $y = 9$.

b Write about what you notice.

3 a Construct tables and on grid 3 plot graphs for: $x = 2$, $x = 4$, $x = 6$ and $x = 8$.

b Write about what you notice.

Remember Extend the lines.

4 a Construct tables and on grid 4 plot graphs for: $y = 10 - x$, $y = 6 - x$ and $y = 2 - x$.

b Describe the slope and direction of the straight lines.

5 Use a computer program, for example, Omnigraph or a graphic calculator to check your answers to questions 1 and 4.

Using conversion graphs

You need:
- a copy of RCM 34
- a ruler
- graph paper

1 This function machine converts kilometres to miles.

km → × 5 → ÷ 8 → miles

a Use the function machine to construct a table of values.

km	8	16	40	80
miles				

b Use your table to plot points on the squared grid. Draw a conversion graph. Give your graph a title.

c Use your graph to convert these distances. Copy and complete the table.

From Glasgow	To Dundee	To Edinburgh	To Gourock	To Largs	To Oban	To Perth
miles	80	45		30	90	55
km			40			

2 This function machine converts °C to °F.

°C → × 9 → ÷ 5 → + 32 → °F

a Use the function machine to construct a table of values. Round up to the nearest whole degree.

°C	0°	5°	16°	28°	35°
°F					

b Plot the points on the graph paper. Draw a conversion graph. Give your graph a title.

c Use your graph to convert these temperatures from °C to °F.

World Temperatures for 31 May

Boston	25°C	Dubai	39°C
Canberra	11°C	Luxor	41°C
Madrid	27°C	Paris	19°C
Tokyo	22°C	Bangkok	33°C

3 4·5 litres ≈ 8 pints.

a Decide how many points you need to plot to draw an accurate graph to convert up to 50 litres to pints.

b Choose suitable scales for the axes.

c Draw the graph on graph paper.

d Label the axes and give the graph a title.

You may need to construct a table of values.

Interpreting and plotting graphs

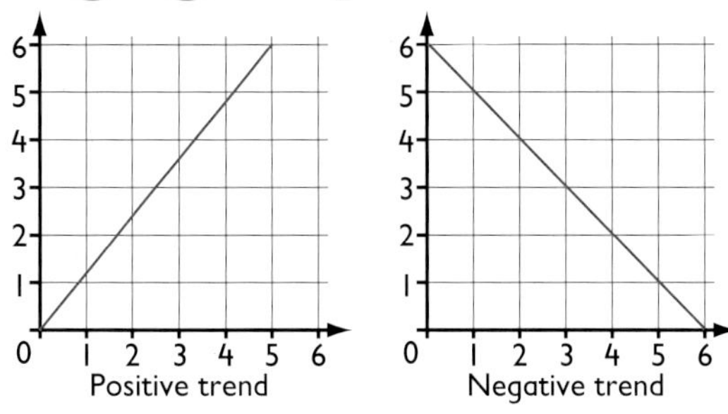

1 For each graph:

a Describe the trend

b Give the value for y when the value for x is:

i 0 **ii** 5 **iii** 10

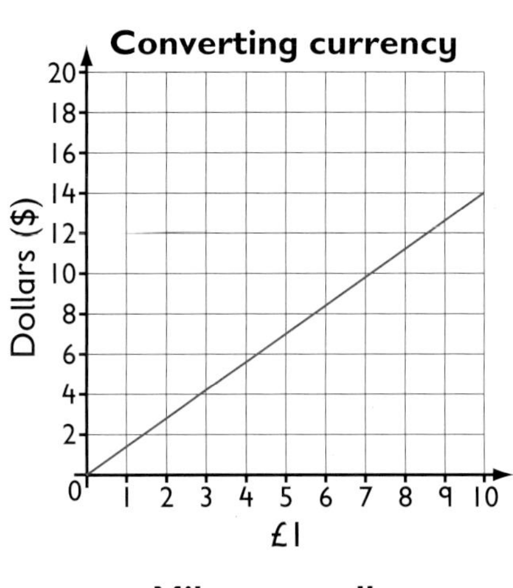

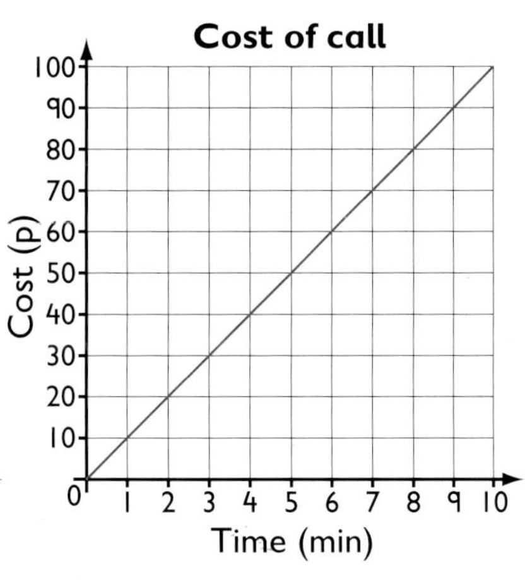

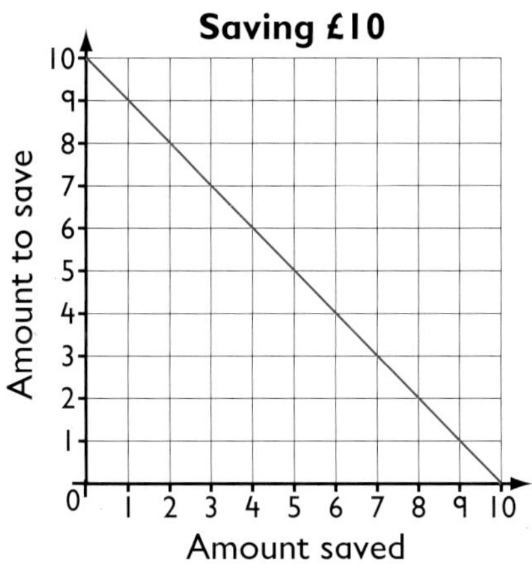

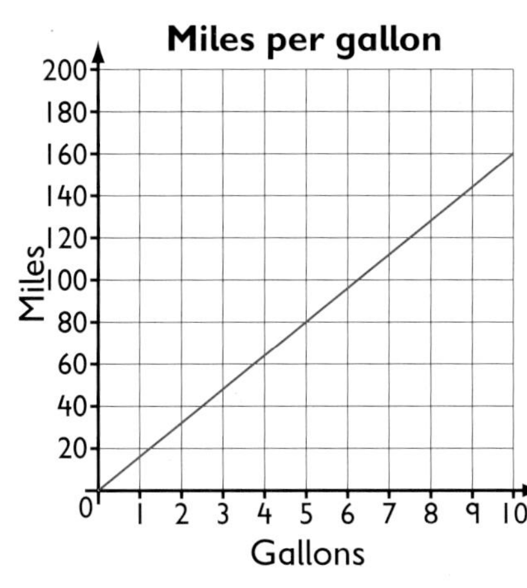

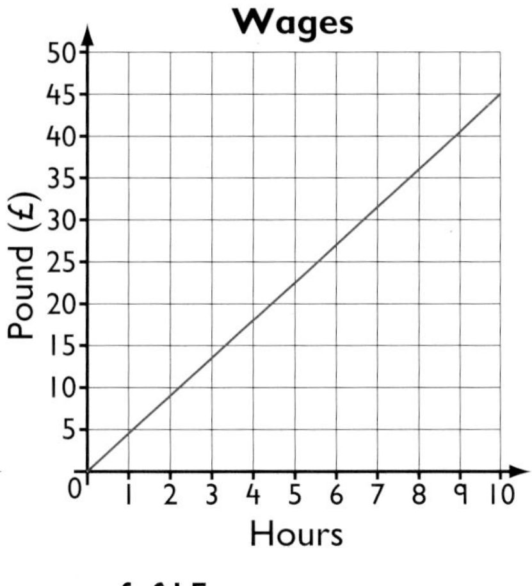

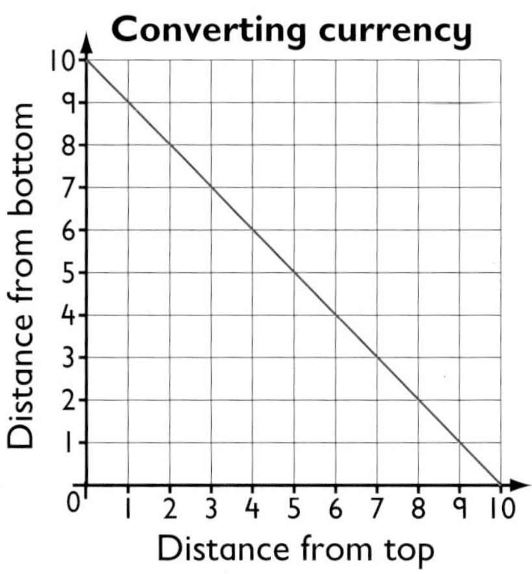

2 The graph shows a fixed charge of £15 rental plus 10p per minute for calls.

Draw graphs of charges for:

a Sports Club charges: £5 fee plus £2 per visit.

b Plumber's charges: £8 call-out fee plus £4 per hour.

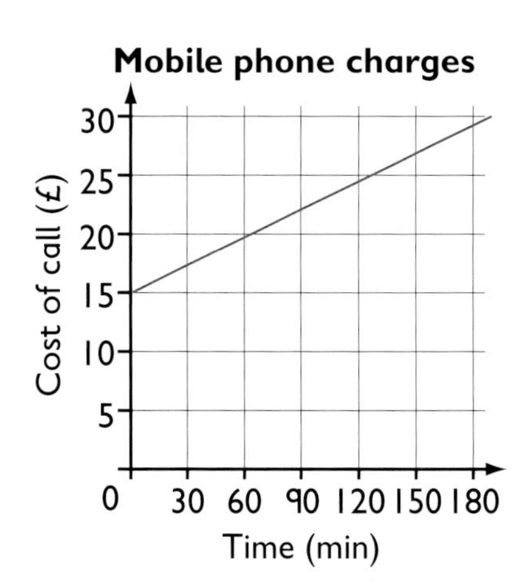

You need:
- square grid paper

Naming lines, sides and angles

1 Name 3 angles that have:

- AB as an arm;
- D as a vertex.

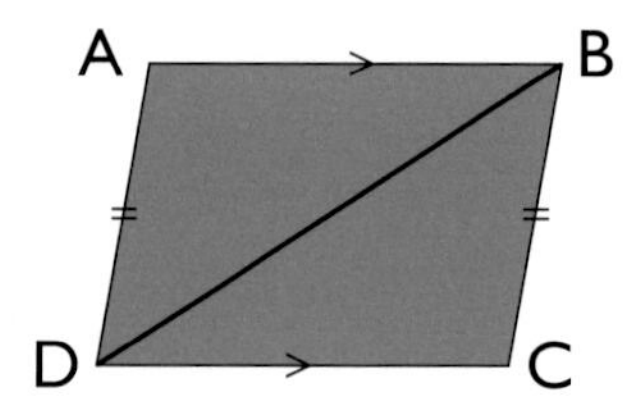

You need:
- a ruler
- a set square
- a protractor

2 Name the equal and parallel sides in these diagrams.

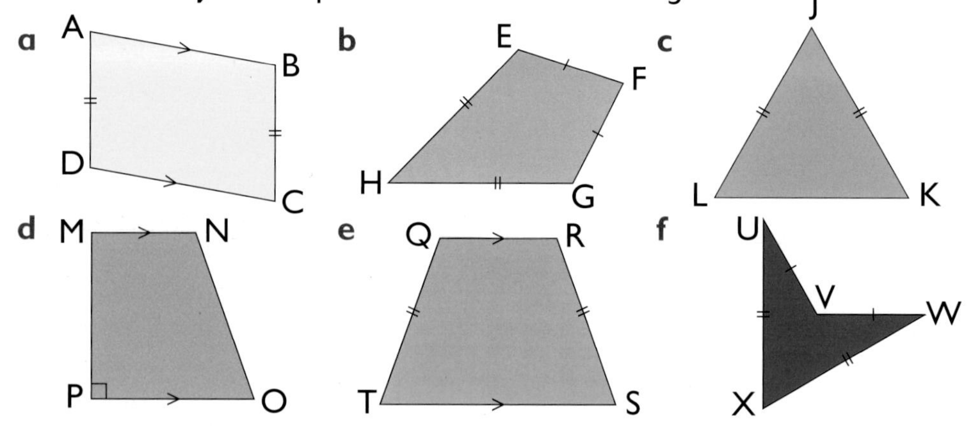

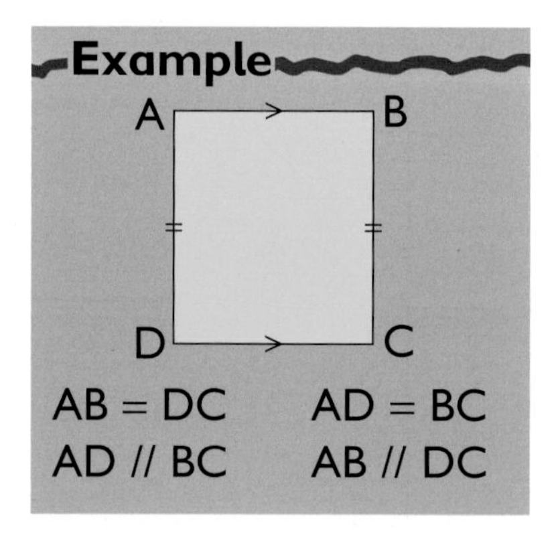

3 Write the name and size of each angle marked with an arc.

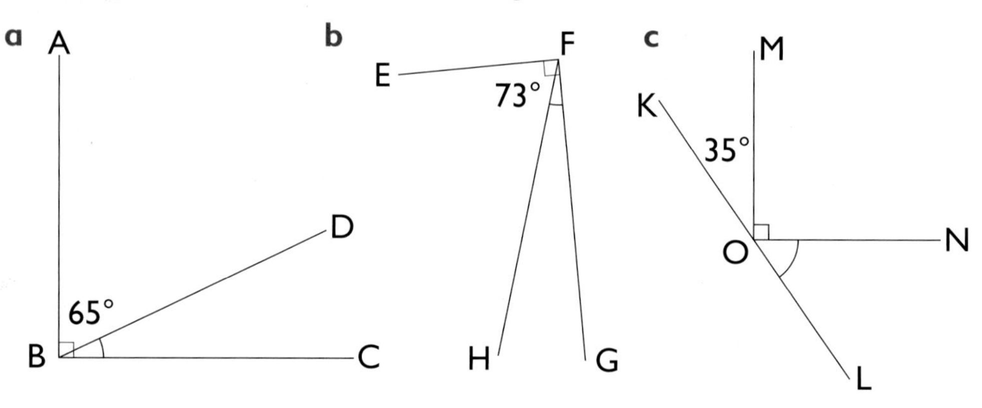

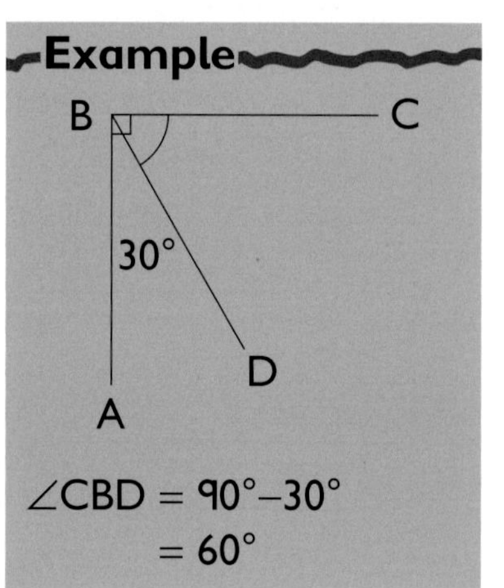

4 a Use your ruler and set square to draw these diagrams.

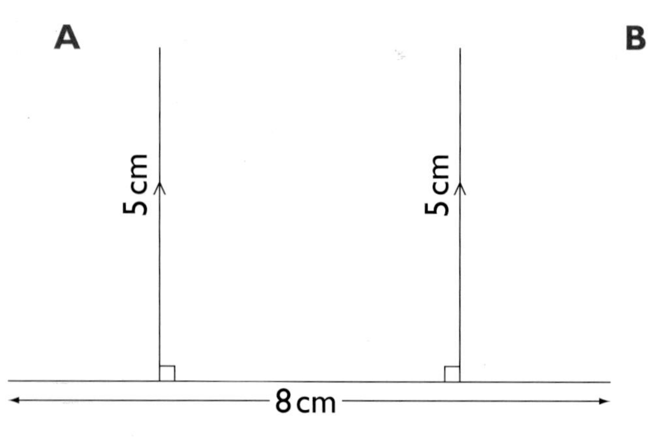

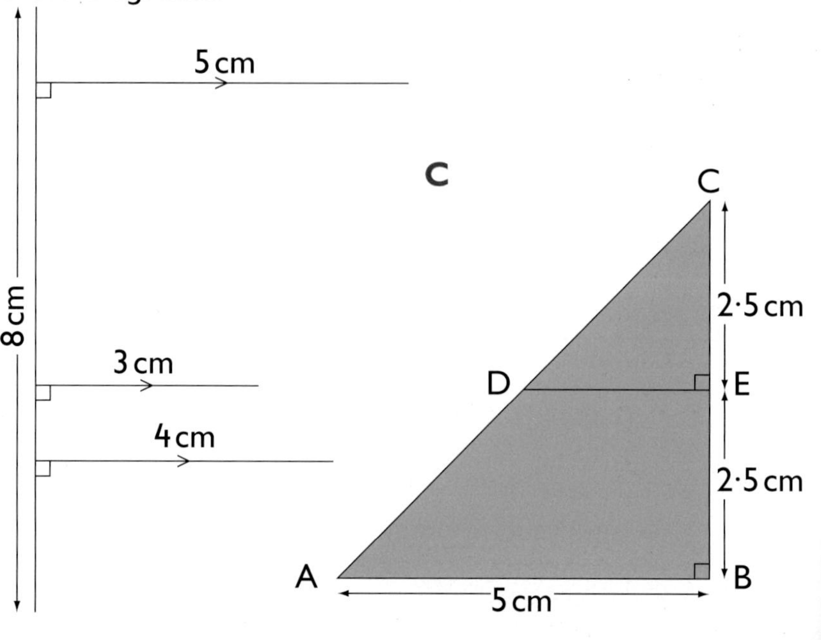

b For the triangle you have drawn:

- measure AD and DC;
- measure ∠BAC, ∠EDC and ∠ACB;
- write about what you find out.

Investigating triangles and quadrilaterals

Checklist	Action
a Mark: equal sides	=
equal angles	∠
right angles	⊾
parallel sides	//
b Colour: parallel sides	blue
perpendicular sides	red
c Draw: lines of symmetry	-------

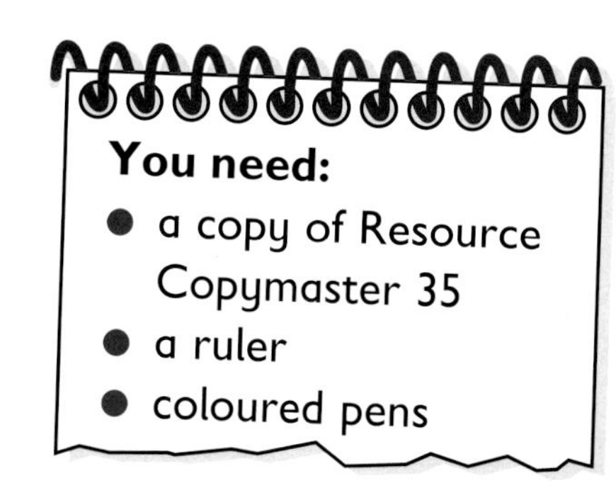

Example

1 Use the 3 × 3 pinboard on Resource Copymaster 35.

  a Construct 8 different triangles.
  Reflections, rotations or translations are not allowed.

  b Use the Checklist to classify the triangles.
  Copy and complete the table.

Property	a	b	c	d	e	f	g
isosceles Δ	✓						
right-angled Δ	✓						
scalene Δ	✗						
line of symmetry	✓						

2 Use the 6 dot pinboards.

  a Draw 4 different triangles.

  b Draw these quadrilaterals: rectangle, rhombus, trapezium, kite, arrowhead.

  c Use the Checklist to classify the shapes.

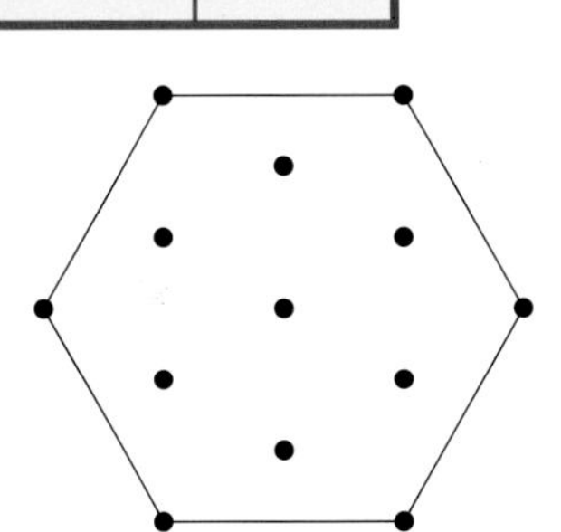

3 Use the 10 dot pinboards.

  a Investigate the different trapezia you can make on the 10 dot board.

  b Use the Checklist to classify the trapezia.

Vertically opposite angles

When two straight lines intersect, the vertically opposite angles are equal.

Thales (c636–546BC)
Founder: 1st School of Greek Mathematics

You need:
- a ruler
- a protractor

1 Measure the angles marked with an arc.

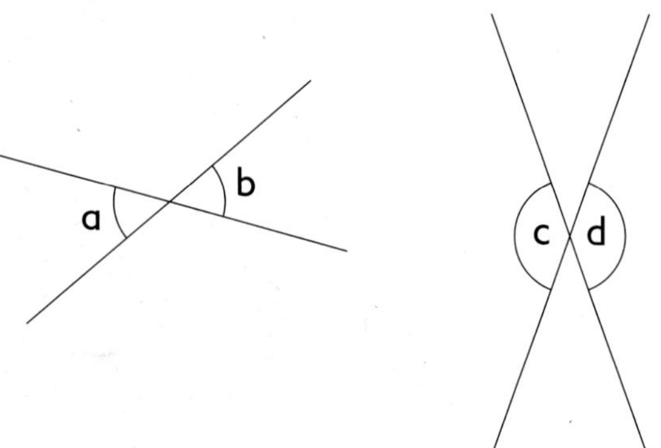

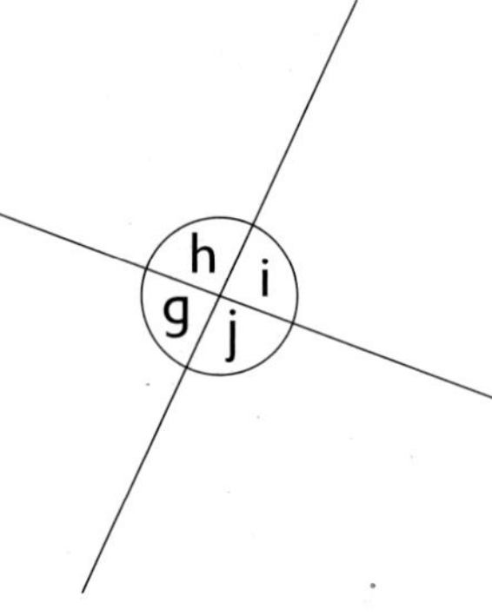

2 Calculate the size of the marked angles.

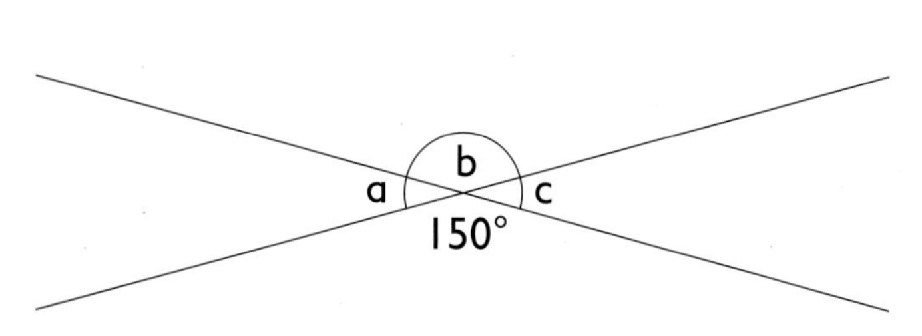

3

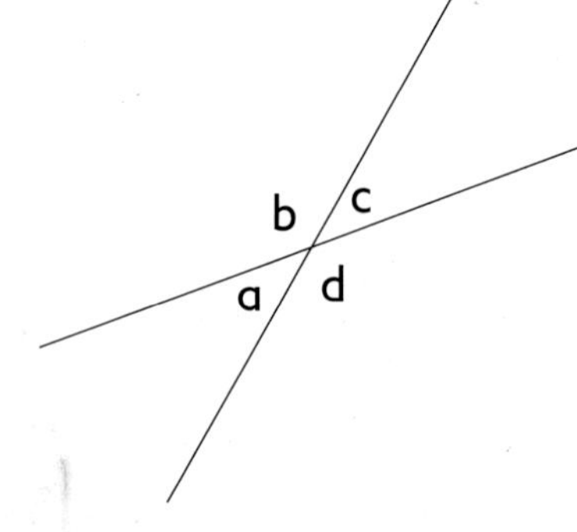

Copy and complete the table for these values of ∠a.

∠a	∠b	∠c	∠d	Total
60°				
55°				
48°				
x°				

4 Use the information in the diagram to find the sizes of the unknown angles.

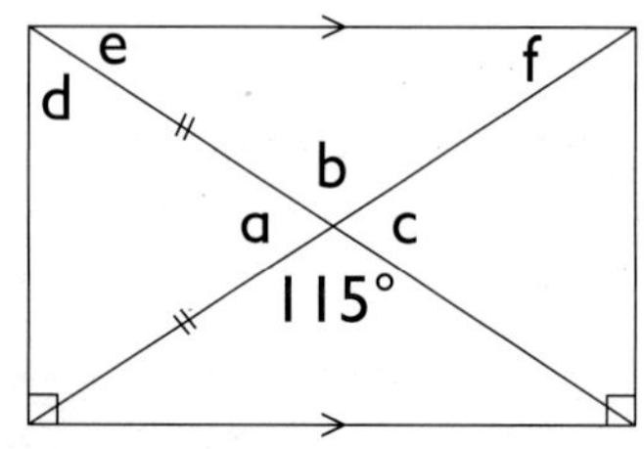

Reasoning about lines and angles

I discovered that the base angles of an isosceles triangle are always equal.

You need:
- a sheet of A4 paper
- scissors
- ruler

1 Find the missing angles in these isosceles triangles.

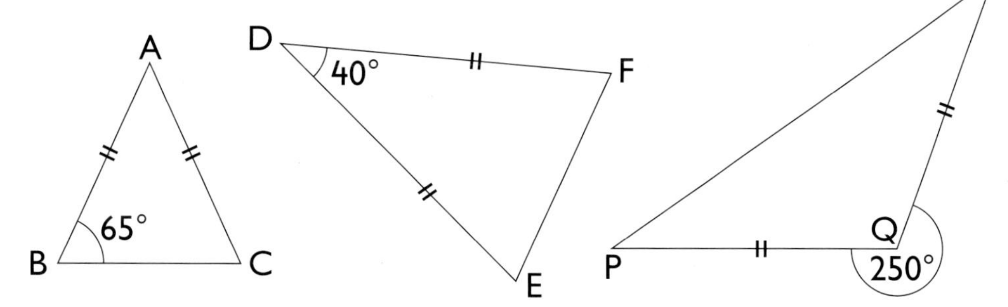

Example

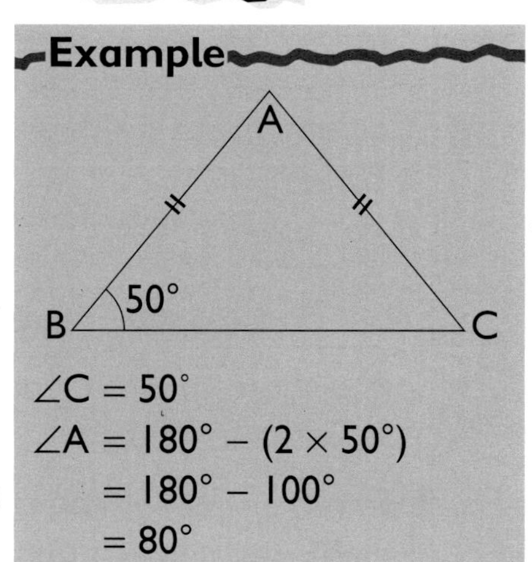

$\angle C = 50°$
$\angle A = 180° - (2 \times 50°)$
$= 180° - 100°$
$= 80°$

2 Use what you know about vertically opposite angles to deduce the size of the angles in each triangle. Check that their sum is 180°.

a 75° K L M 80° 25°

b 30° R T 75° S

3
- Fold a sheet of A4 paper twice to make 4 layers.
- Draw any quadrilateral.
- Cut out through all 4 layers to make 4 identical quadrilaterals.
- Fit the 4 quadrilaterals round a point.
- Copy and complete. The 4 angles add to _____ right angles or _____°.

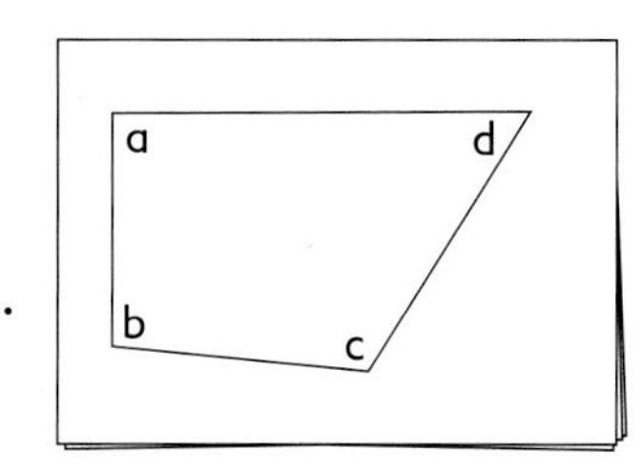

4 Calculate the missing angles in these quadrilaterals.

a

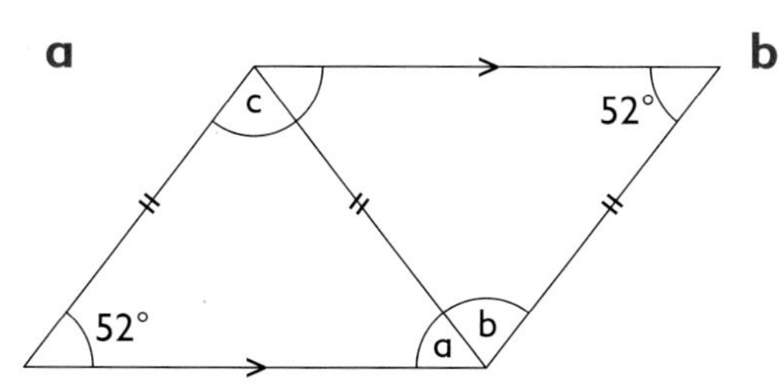

b

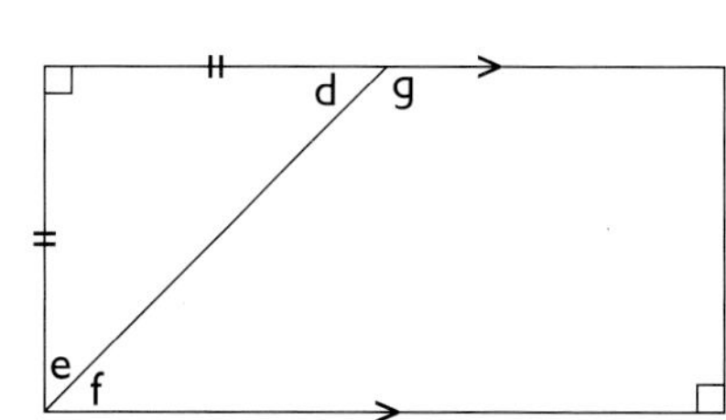

c

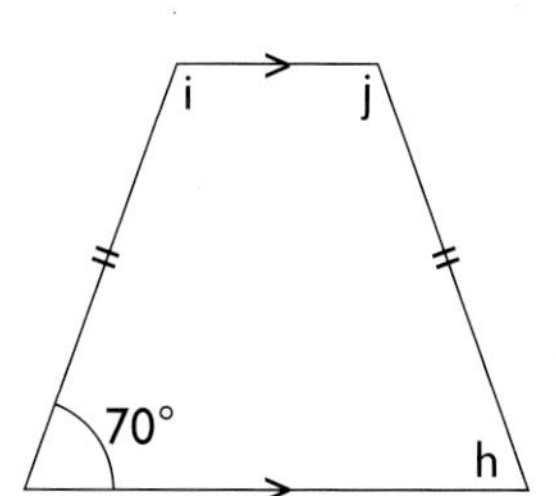

5 Calculate the size of the marked angles in the tiling patterns.

Investigating shapes in overlaps

You need:
- 2 congruent squares
- Resource Copymaster 37
- 1 cm triangular dot paper
- a ruler

1 What different shapes can you make by overlapping 2 squares?

- Use your squares as templates to sketch the shapes you make.
- Name the shape in the overlap.
- Try to find these shapes: square, isosceles triangle, kite, pentagon, hexagon, heptagon, octagon.
- Can you make a rhombus or a trapezium? If not, explain why.

Example

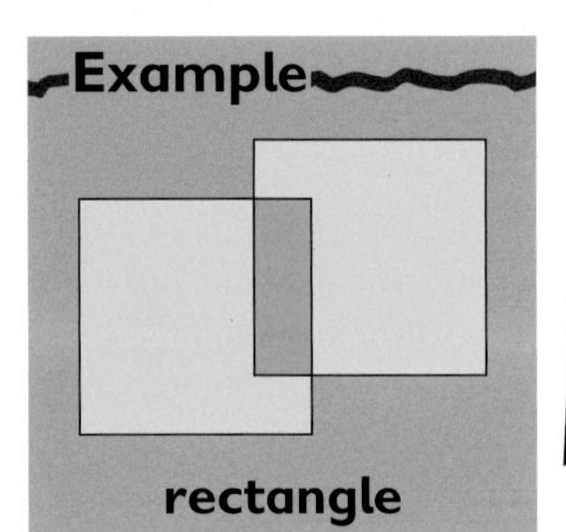

rectangle

2 On Resource Copymaster 37: Overlapping quadrilaterals:

a make these shapes by overlapping 2 congruent trapezia;

b name the shape in the overlap each time: square, rectangle, parallelogram, isosceles trapezium, pentagon.

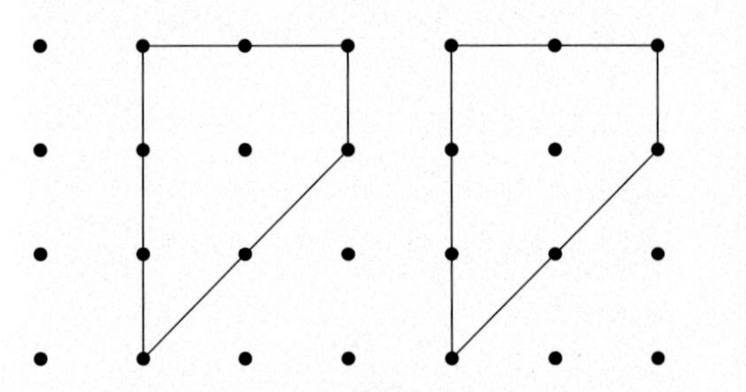

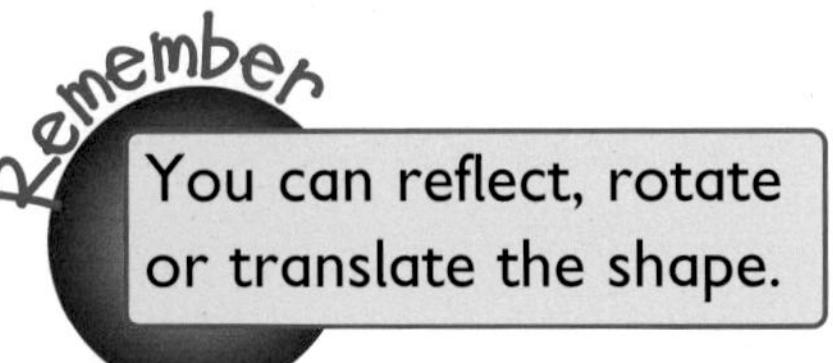

3 On Resource Copymaster 37, Overlapping quadrilaterals: investigate and name the shapes you can make by overlapping 2 congruent parallelograms.

4 Investigate the different overlapping shapes you can make with an equilateral triangle and an isosceles trapezium. Record on 1 cm triangular dot paper. Shade the shape in the overlap and name it.

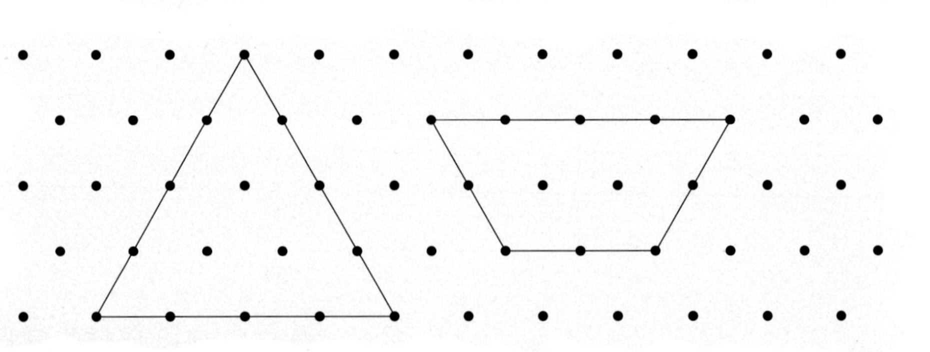

Here are some ideas: 2 different equilateral triangles, rhombus, pentagon
2 different parallelograms, 3 different trapezia, hexagon

Drawing 3-D shapes

You need:
- Resource Copymaster 38
- interlocking cubes
- ruler

1 On Resource Copymaster 38 record all the possible solids that can be made from 4 cubes.

Make the shapes to check your answers.

2 Each of these 3-D shapes is made with 6 cubes. Work out which drawings are different views of the same shape.

a

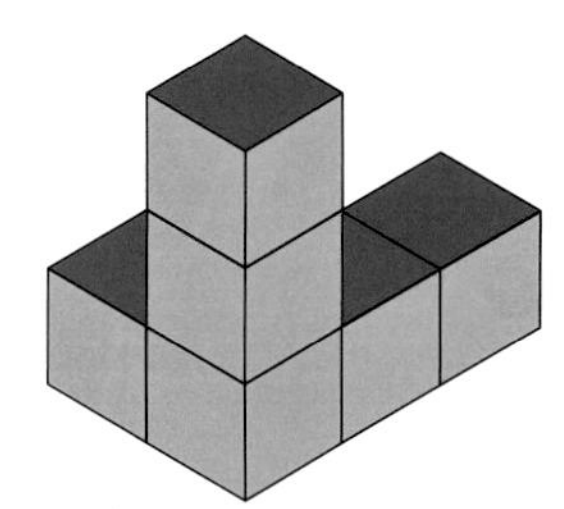

b

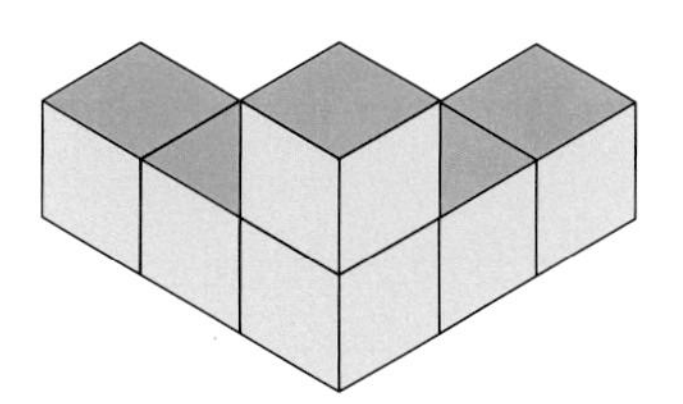

c

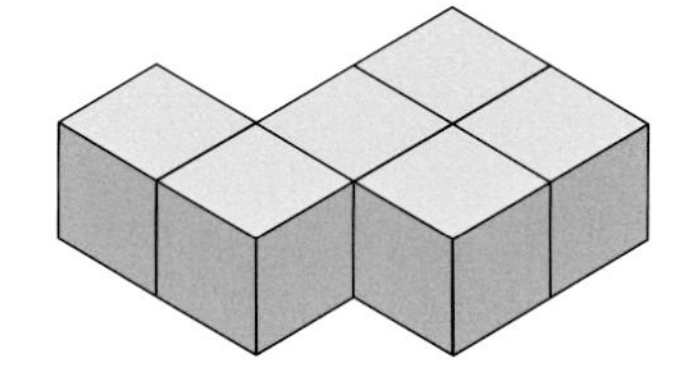

d

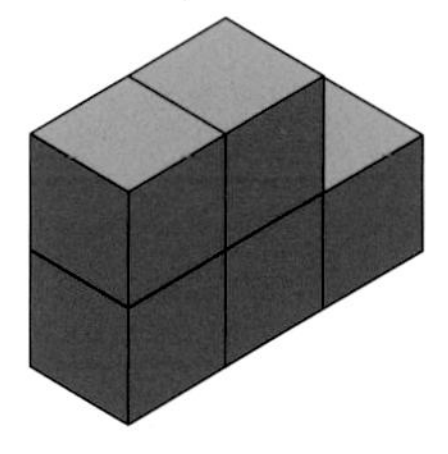

e

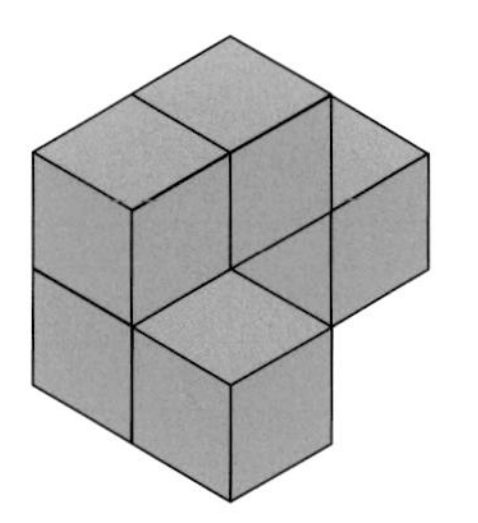

f

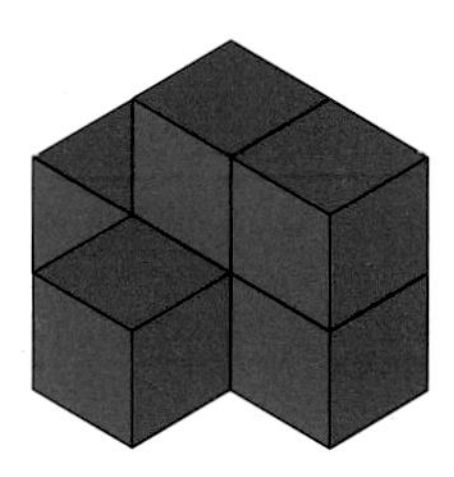

g

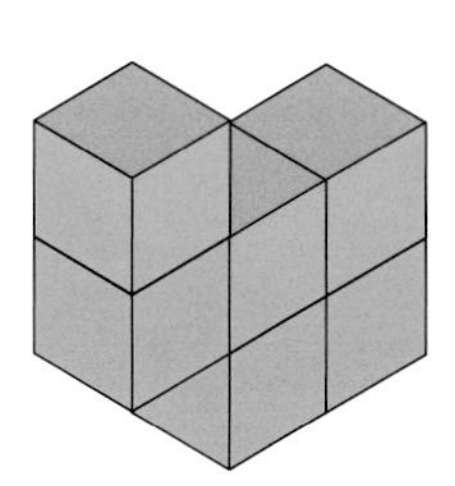

h

i

3 **a** Build the next skeletal cube in the pattern.

b Copy and complete the table.

Size of cube	$3 \times 3 \times 3$	$4 \times 4 \times 4$	$5 \times 5 \times 5$	$6 \times 6 \times 6$
Number of cubes				
Difference				

c Use the difference pattern to predict the number of cubes for a $7 \times 7 \times 7$ skeletal cube.

d What if… you had a $10 \times 10 \times 10$ skeletal cube?

Skeletal cubes

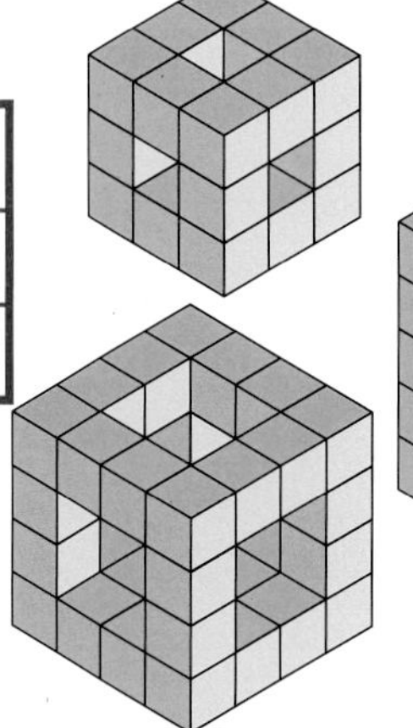

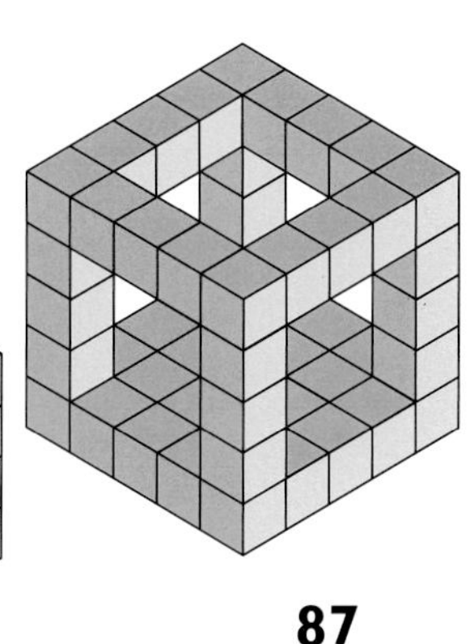

Investigating packing case moves

In the warehouse, each packing case can be moved on to an adjacent empty space. The fork-lift truck driver can move a packing case in a direction either parallel or perpendicular to the other cases. He cannot move a packing case diagonally.

You need:
- 1 red counter
- 14 blue counters
- large square grid paper

1 In this 3 × 3 grid section of the warehouse:

Find the fewest number of moves for the red packing case to reach the empty space.

Hint: Begin by investigating for a 2 × 2 grid.

2 Describe the route that gives the least number of moves.

3 What if the warehouse has a 4 × 4 grid section?
Find the least number of moves to place the red packing case in the empty space.

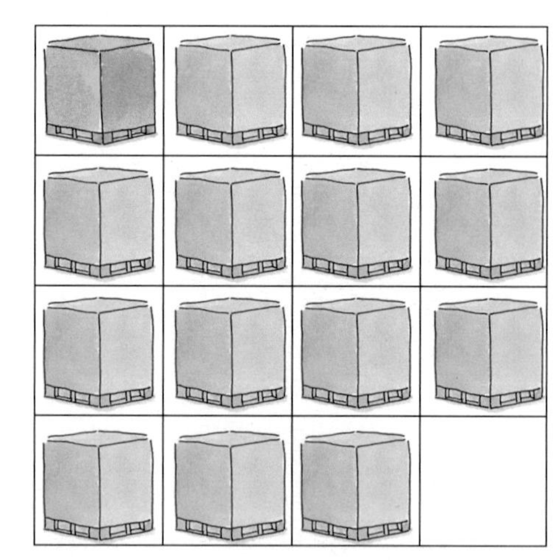

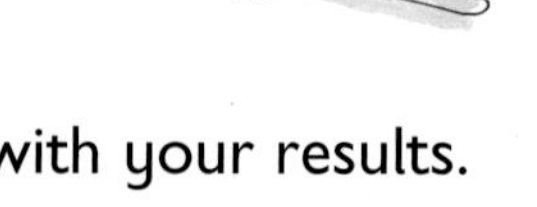

4 **a** Complete the table with your results.

Side of grid section	2 × 2	3 × 3	4 × 4	5 × 5	6 × 6
Least number of moves	5				
Difference					

b Use the difference pattern to predict for a 5 × 5 and a 6 × 6 grid.

c N represents the least number of moves and S the side of the grid. Write a formula to show the relationship between N and S.

N = ____________

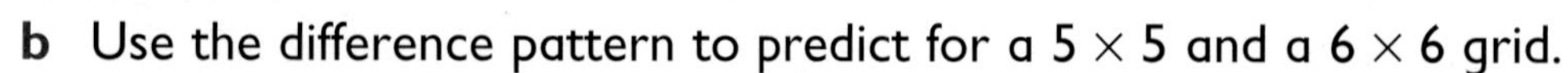

d Check that your formula works.

Predicting reflections

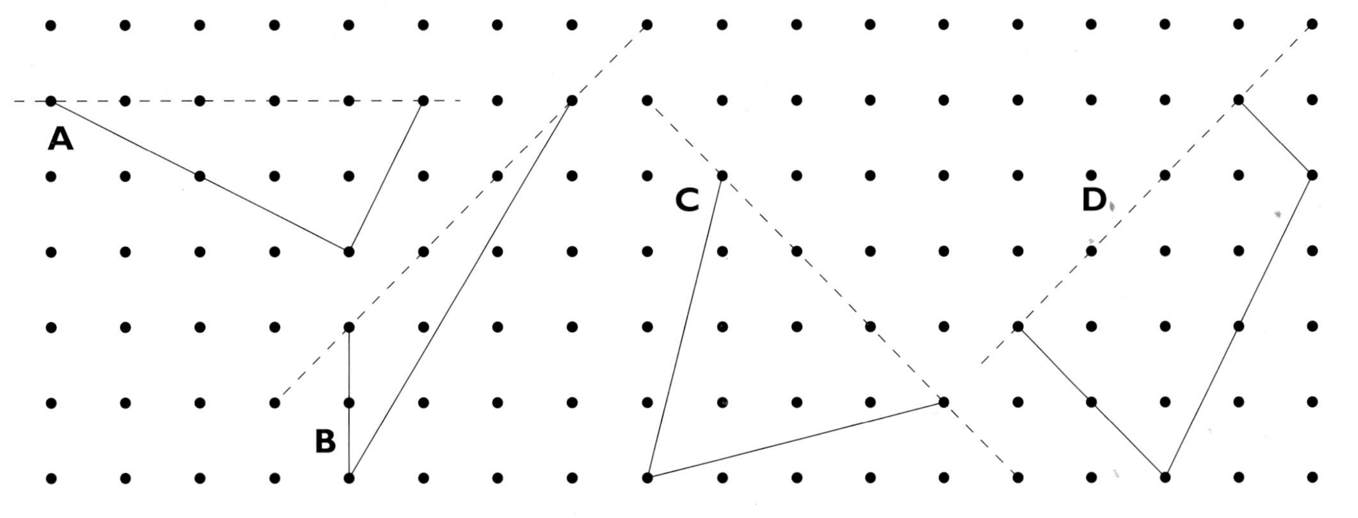

1 a Copy each shape on to 1 cm square dot paper.
b Reflect each shape in the mirror line.
c Name the shape you make.
d Mark the equal sides and equal angles.

You need:
- 1 cm square dot paper
- a ruler
- coloured pens

2 Reflect each word in horizontal, vertical and sloping mirror lines.

PETS **BIKE** **TWIN**

Example

FANS

3 Copy each shape on to 1 cm square dot paper.
For each shape, reflect it both ways in the mirror line.

Check that you have made a symmetrical pattern each time.

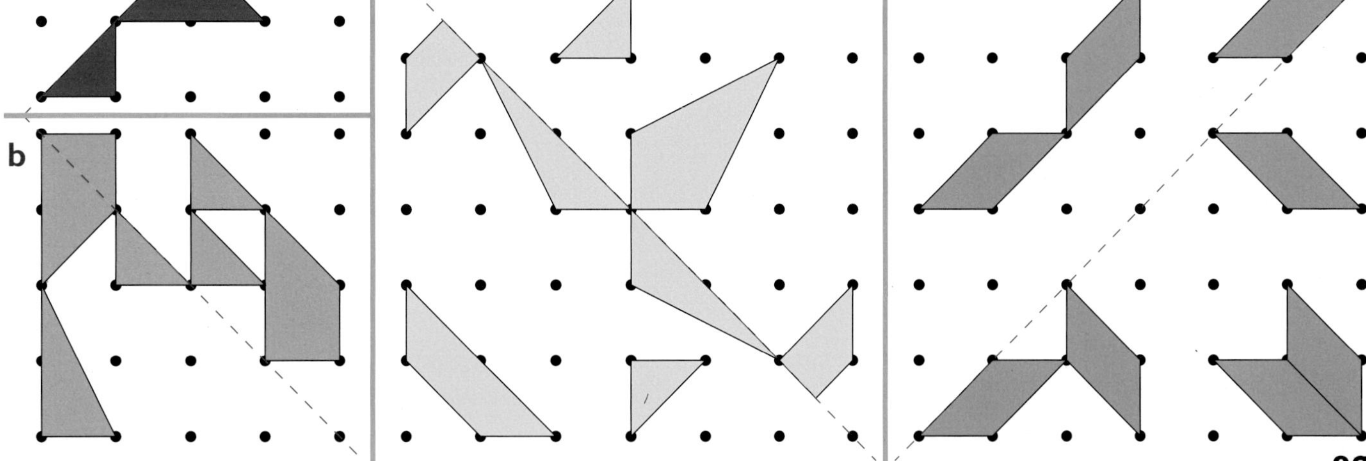

Investigating reflections

You need:
- 10 interlocking square tiles
- 1 cm square grid paper

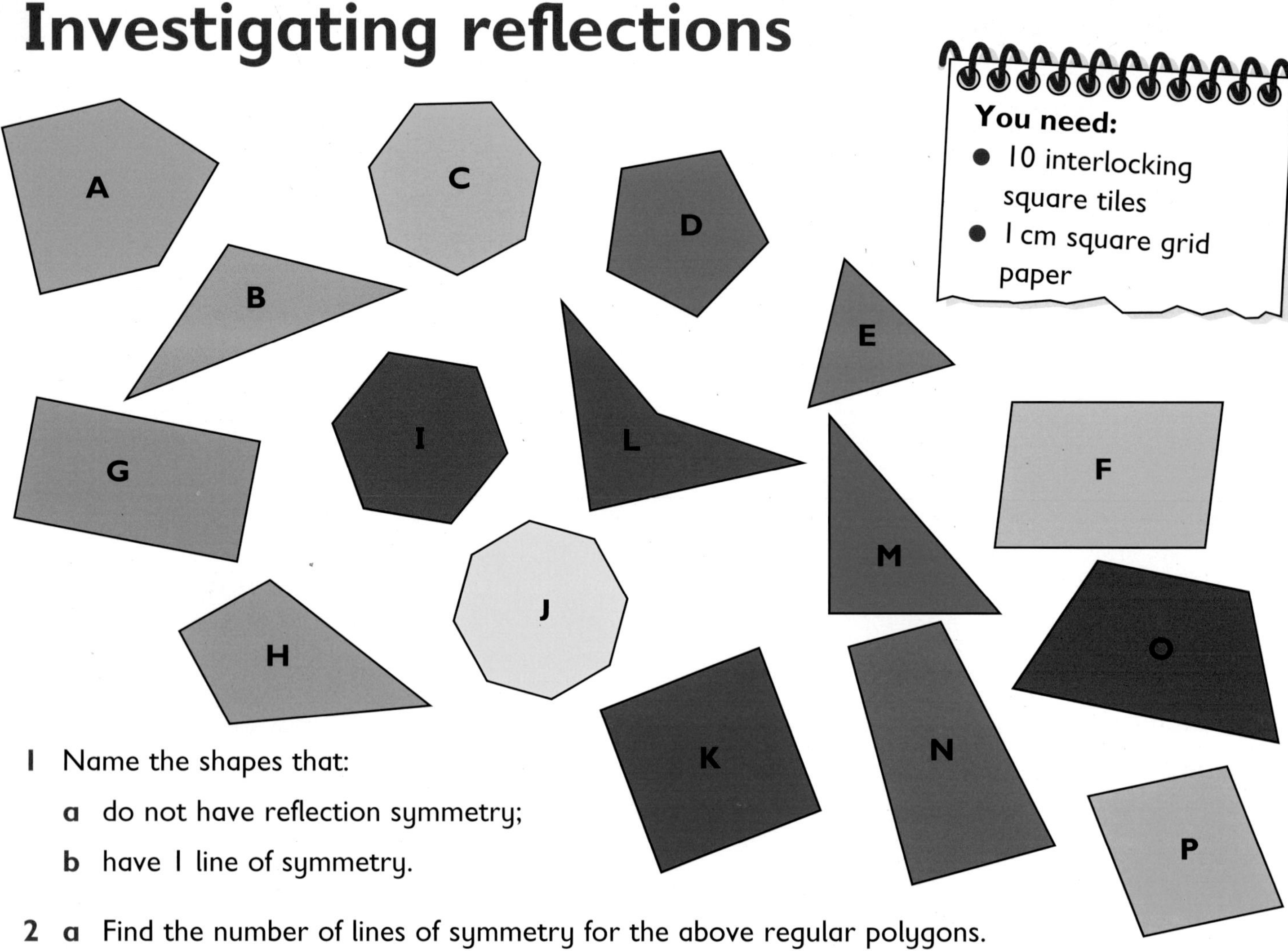

1 Name the shapes that:

a do not have reflection symmetry;

b have 1 line of symmetry.

2 a Find the number of lines of symmetry for the above regular polygons.

b Write your answers in a table.

c Use the table to predict for regular polygons of 10 sides, 12 sides and 20 sides.

Number of sides (S)	3	4	5	6	7	8	9	10	12	20
Number of lines of symmetry (L)										

d Write a formula that shows the relationship between S (number of sides of a regular polygon) and L (number of lines of symmetry).

3 Make the pentominoes with your interlocking square tiles.

For each pair of pentominoes, investigate ways to fit them together to make a shape that has reflection symmetry.

Record your findings on 1 cm squared grid paper.

a

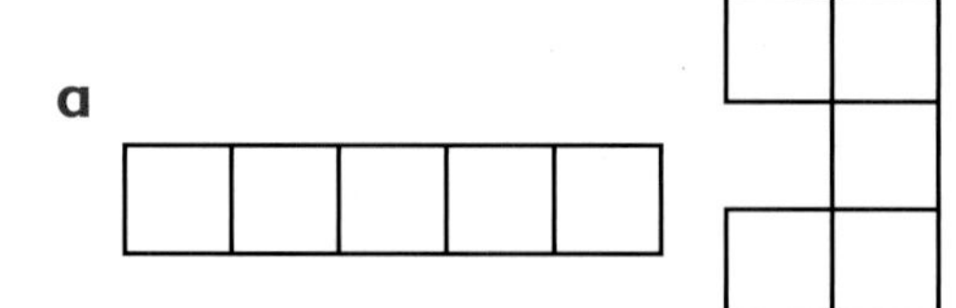

b c

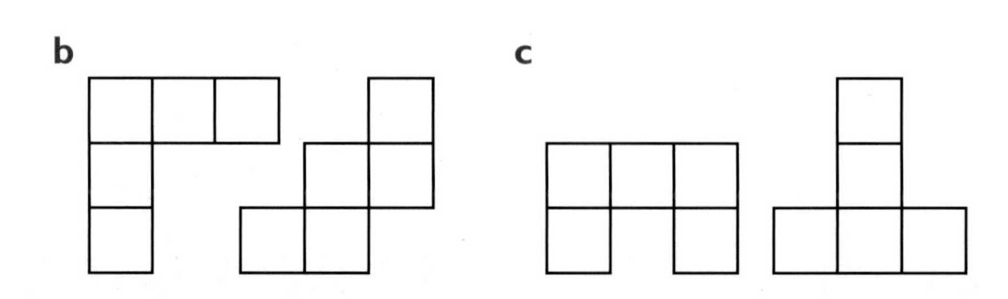

Centres of rotation

1 For each shape, describe the rotation of shape A to shape Á about the centre of rotation.

a

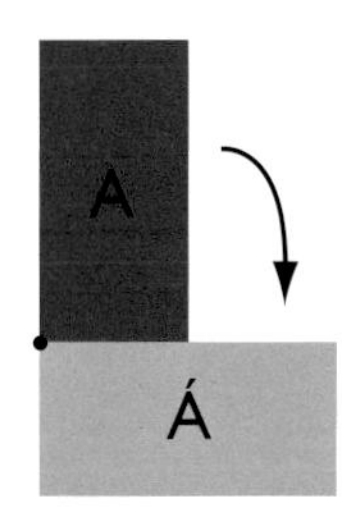

b

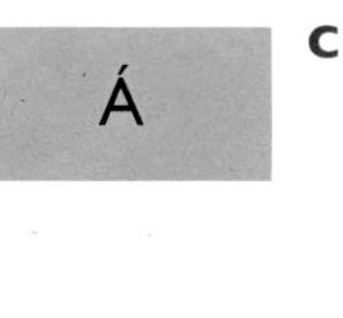

c

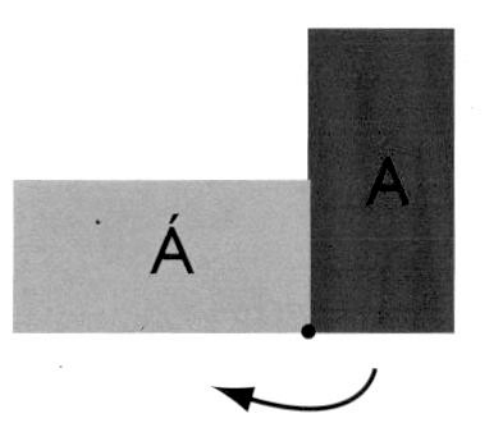

d

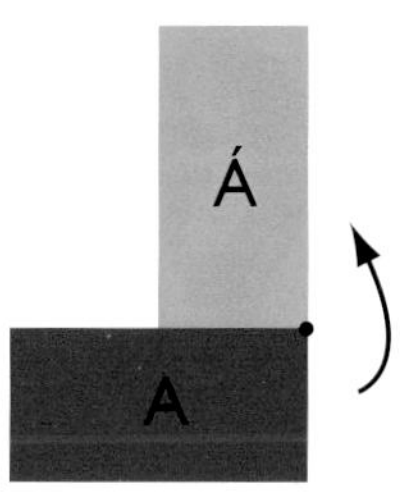

2 Copy these diagrams on to squared paper and draw the completed shape.

a Rotate clockwise through 90° about the dot.

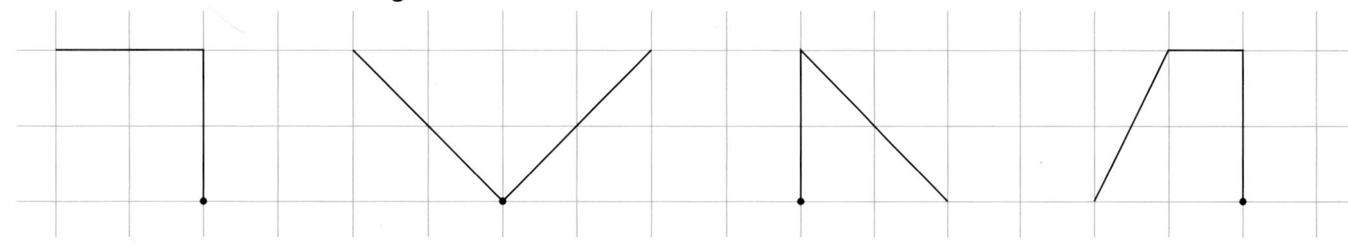

b Rotate through 180° about the dot.

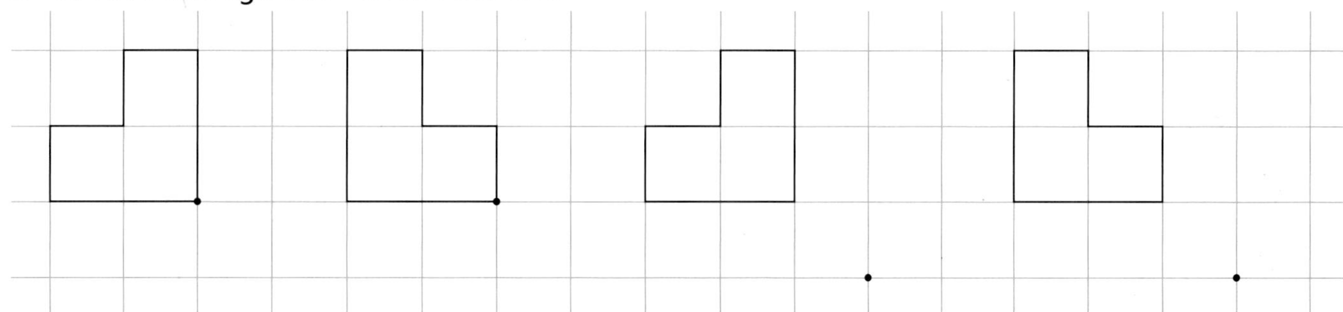

3 a On your card, draw a square 4 cm × 4 cm and mark the mid point of each side.

b Use the 1p coin to draw the semicircle.

c Draw the rotation of the semicircle through 90°.

d Repeat for the diagonally opposite vertex.

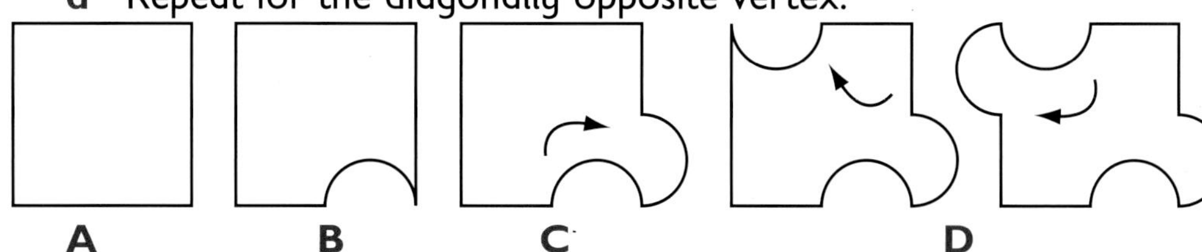

A **B** **C** **D**

You need:
- card
- 1p coin
- scissors
- ruler
- colouring materials

e Make a tiling pattern by rotating your card template.

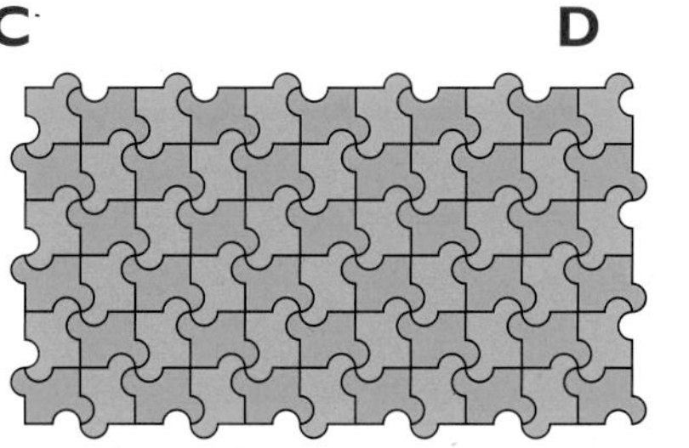

Order of rotational symmetry

1 Write the order of rotational symmetry for these designs.

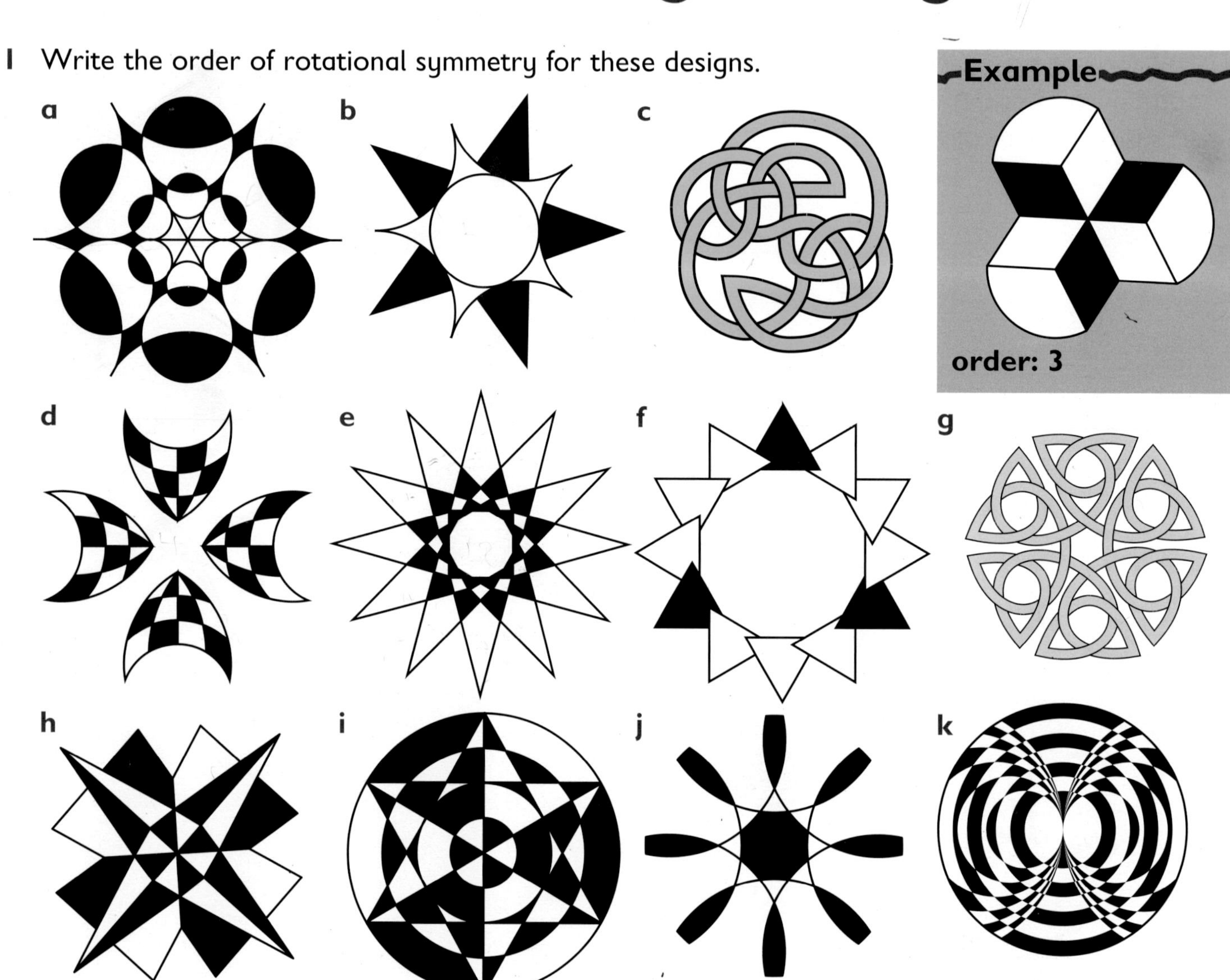

2 Write the letter of each design that has reflection symmetry.

3 Use a different grid on your Resource Copymaster 39 of 4 quadrant co-ordinate grid paper for each question.
Draw the original parallelogram on each grid.
Draw the position of the parallelogram after:

a 180° rotation about S;

b 90° anticlockwise rotation about S;

c 180° rotation about P;

d 90° clockwise rotation about R.

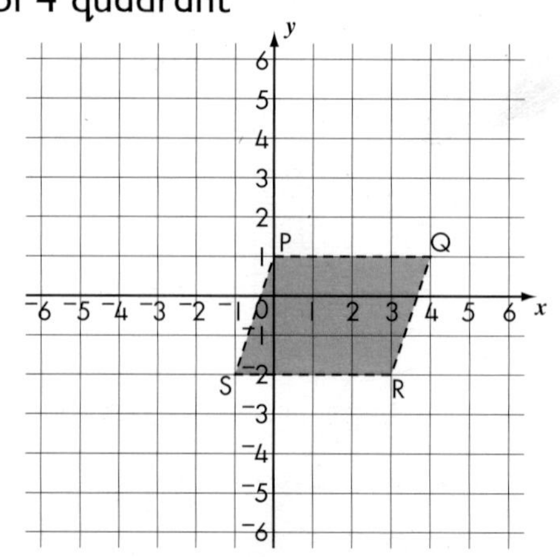

4 Write the co-ordinates of Q after each rotation.

Translating shapes

Remember

The shape translates
- to the left or right
- up or down

1 For each grid, describe the translation of shape A to shape Á.

a

b

c

d

2 Use a different grid on your Resource Copymaster 39 of 4 quadrant co-ordinate grid paper for each question.
Translate the green shape in question 1:

a 2 units left, 3 units down;

b 6 units right, 2 units down;

c 4 units right, 3 units up;

d 1 unit left, 8 units up.

3 For each translation of the blue hexagon:

a write the distance and direction;

b write the new co-ordinates of the circled vertex.

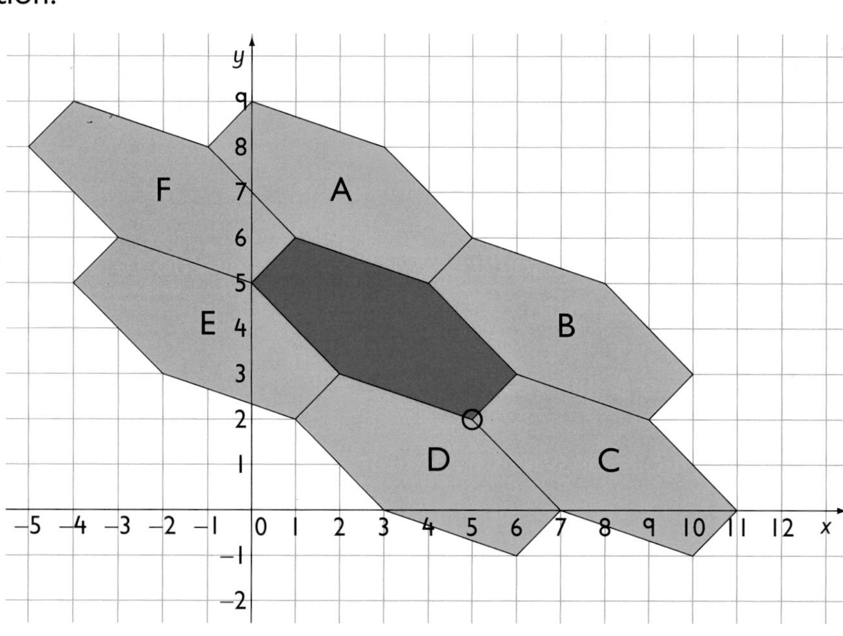

- Recognise and visualise the transformation and symmetry of a 2-D shape:
  - reflection in given mirror lines, and line symmetry;
  - rotation about a given point, and rotational symmetry

Investigating transformations

This shape is called an L-shaped tromino.

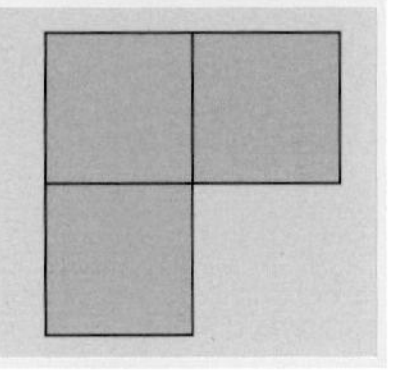

You need:
- 10 interlocking square tiles
- 1 cm squared paper

1 Make 2 L-shaped trominoes with your tiles.
Use 2 L-shaped trominoes touching edge to edge to draw:

   a 2 different shapes, each with only 1 line of symmetry;

   b 2 different shapes, each with rotational symmetry of order 2;

   c a shape with 2 lines of symmetry and rotational symmetry of order 2.

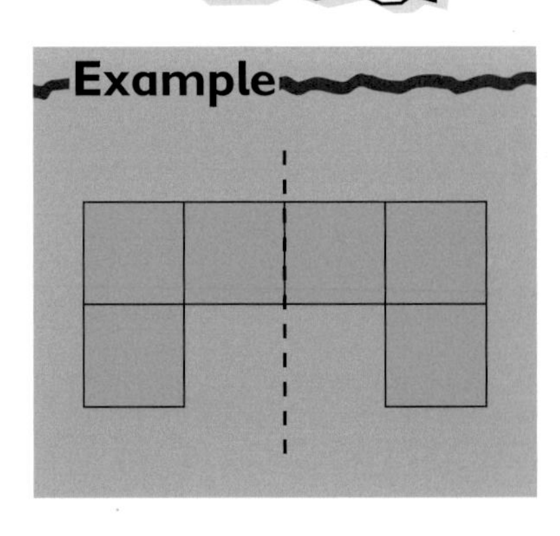

2 Make 2 U-shaped pentominoes with your tiles.
Use 2 U-shaped pentominoes, touching edge to edge to draw:

   a 2 different shapes each with only 1 line of symmetry;

   b 2 different shapes each with rotational symmetry of order 2;

   c a shape with 2 lines of symmetry and rotational symmetry of order 2.

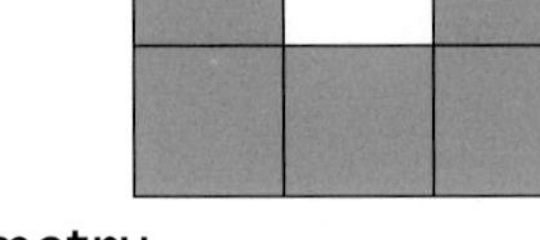

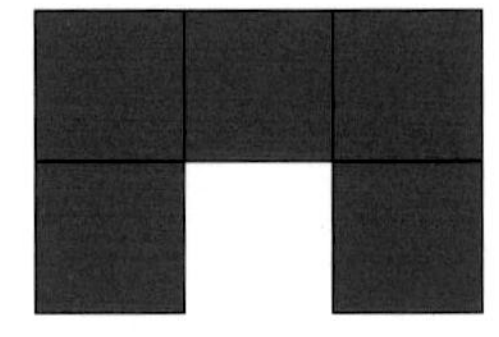

3 Now investigate for these pentominoes.

   a

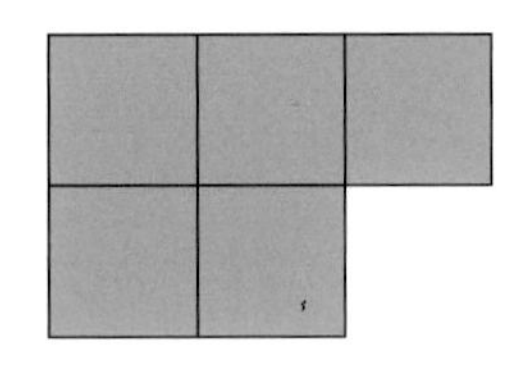

   b

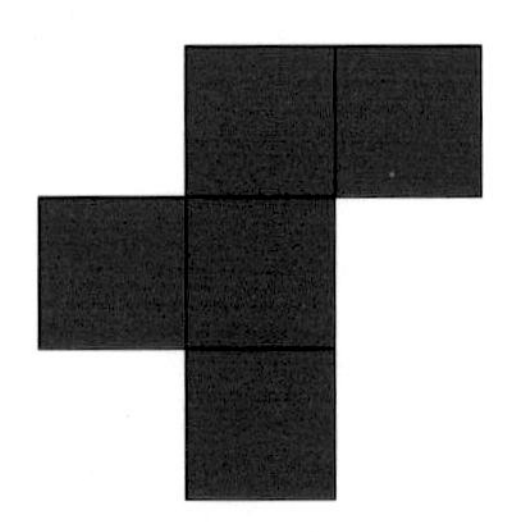

4 Combine these 2 squares with the pentomino to make a shape with reflection symmetry.

Try to find 5 different solutions.

Record your answers on squared paper.

Check your shapes for rotation and mark the centre of rotation with a ●.

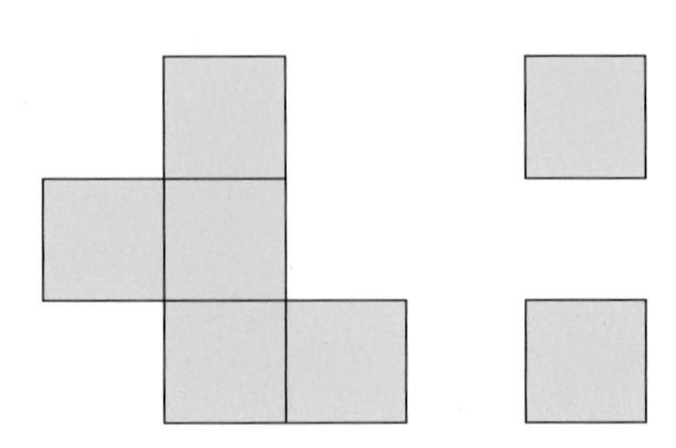

- Recognise and visualise the transformation and symmetry of a 2-D shape:
  – reflection in given mirror lines, and line symmetry;
  – rotation about a given point, and rotation symmetry; – translation

Transformations in a tiling pattern

Colour Key

shape 1 orange
shape 2 blue
shape 3 yellow

You need:
- a copy of Resource Copymaster 40
- coloured pens

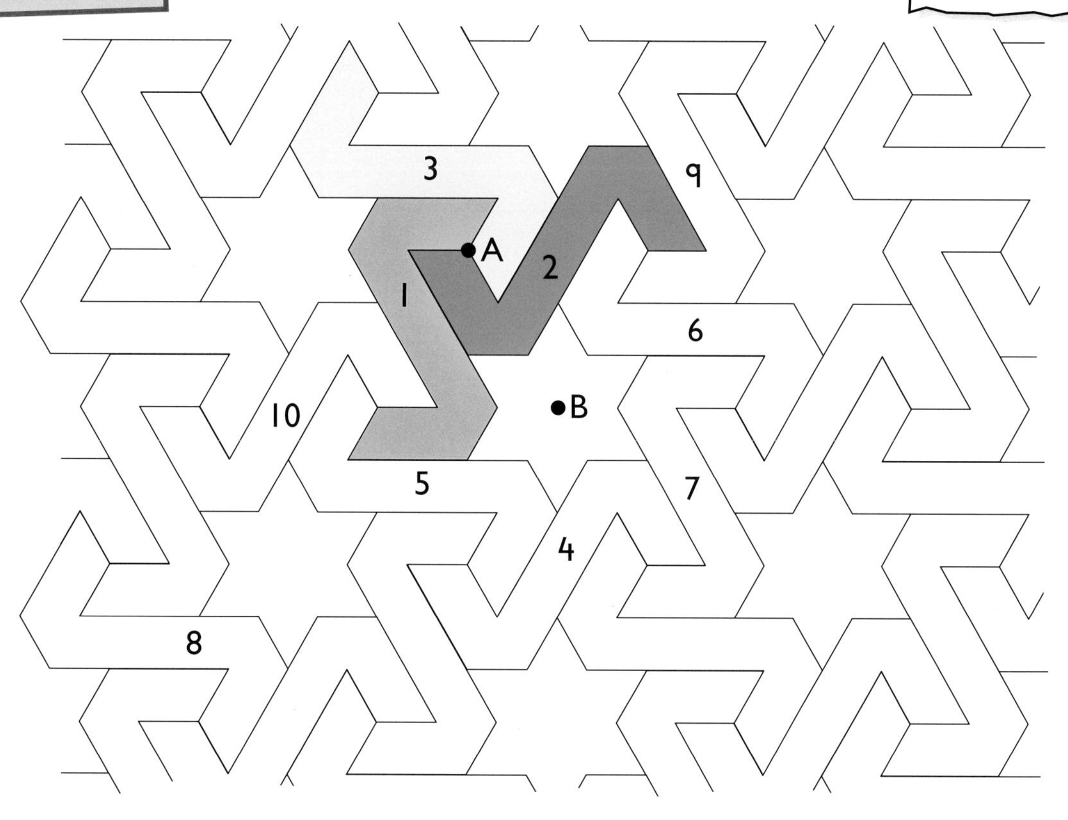

1 **a** On Resource Copymaster 40, colour shapes 1, 2 and 3 as shown.

b Using the colour key, locate and colour two unnumbered translations of shape 1, shape 2 and shape 3.

2 Write the number of the two shapes which are:

a translations of shape 1;

b translations of shape 2;

c translations of shape 3.

3 Write the number of the shape that is:

a a 120° clockwise rotation of shape 1 through point A.

b a 240° anticlockwise rotation of shape 3 through point A.

c a 180° rotation of shape 5 through point B.

4 Use the colouring key to complete the tiling pattern.

Investigating transformations

You need:
- a copy of Resource Copymaster 41
- a ruler
- scissors
- coloured pens

1 On a 3×3 pinboard you can make one translation of this shape.

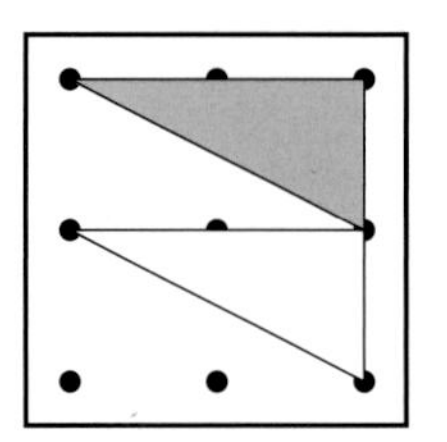

Use Resource Copymaster 41.
Find the number of translations of the shape you can make:

a on a 4×4 pinboard;

b on a 5×5 pinboard.

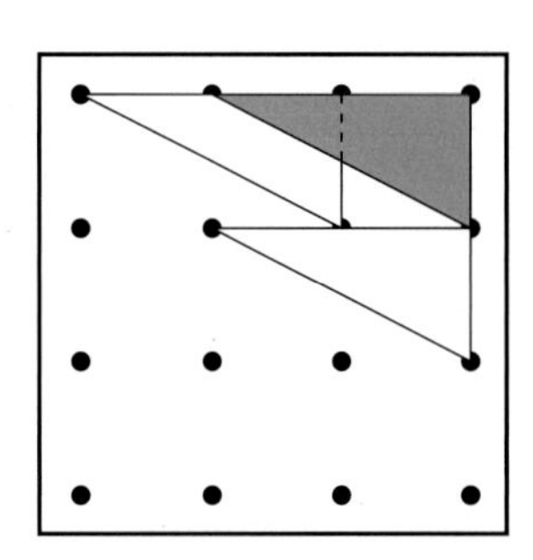

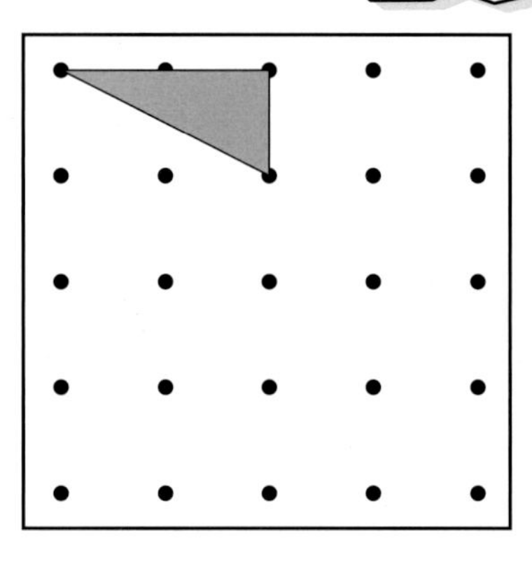

2 Mark a square into equal divisions.

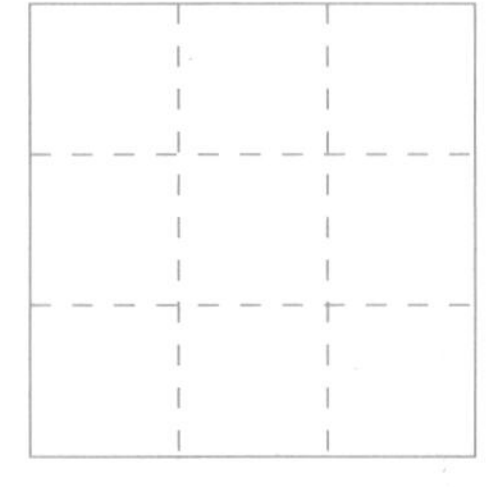

Make a design to meet the sides of the square.

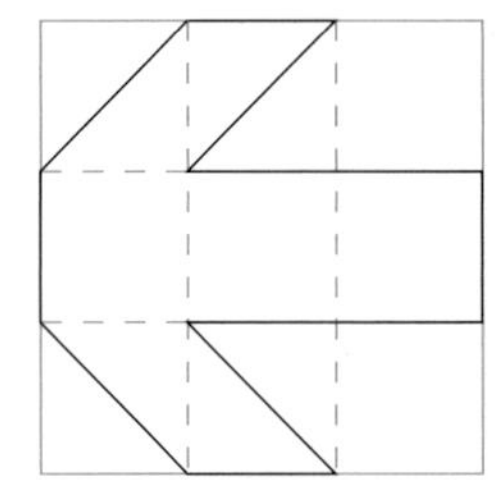

Mark a corner. Rotate the tile to make a pattern.

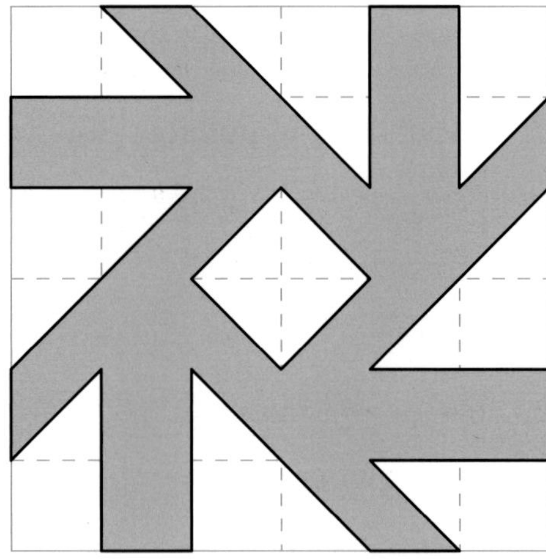

a Copy the design onto all 4 squares at the bottom of the page.
Cut out the square tiles.

b Use the 4 tiles to explore designs you can create by rotating, reflecting or translating the tile. Make a different design in each large square.

c Decide which design you like best.
Use it to tile the rectangular area.
Colour your tiling design.

Co-ordinates of 2-D shapes

You need:
- a copy of Resource Copymaster 39

1 a List the co-ordinates of the four points that are vertices of:

i a square
ii a parallelogram
iii a kite
iv a trapezium

b List the co-ordinates of the vertices of a non-right-angled isosceles triangle.

c The points D (1, −1) and E (−2, −4) are two vertices of a right-angled isosceles triangle. Find two different co-ordinates for the third vertex.

d AE and EF are two sides of a rectangle. Write the co-ordinates of the vertex G.

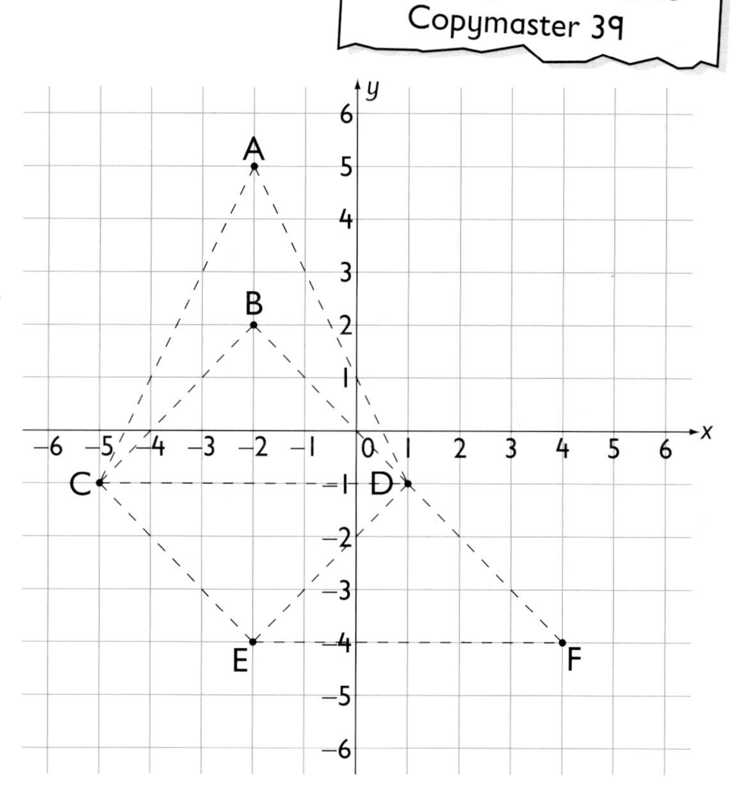

2 Plot these points on Resource Copymaster 39: P (−4, −3), Q (1, −3), R (3, 4).
Write the co-ordinates of the 4th vertex S if the shape is:

a a kite **b** a parallelogram **c** a trapezium

Justify your decisions.

3 Ian has to write a program for a computerised machine to draw the net for a cube without going over the same line twice.

He must write the co-ordinates of the corners in an ordered route.

He makes this sketch of the route.

Start: (−1, −2), (1, −2)

Continue the route and write the co-ordinates to complete the program.

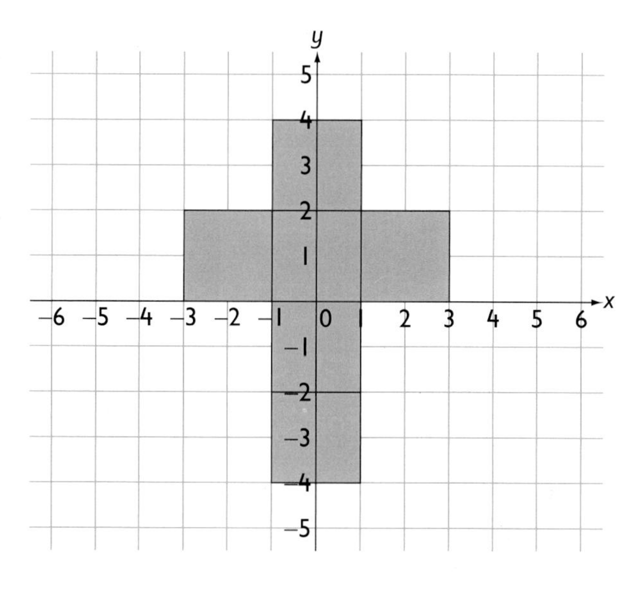

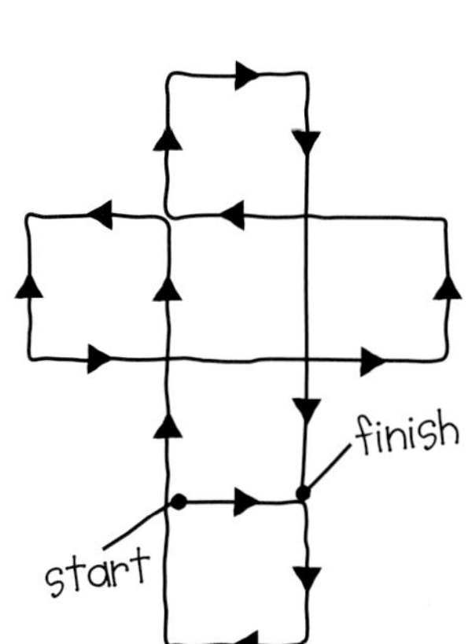

4-quadrant rotations and reflections

You need:
- a copy of Resource Copymaster 39
- a ruler

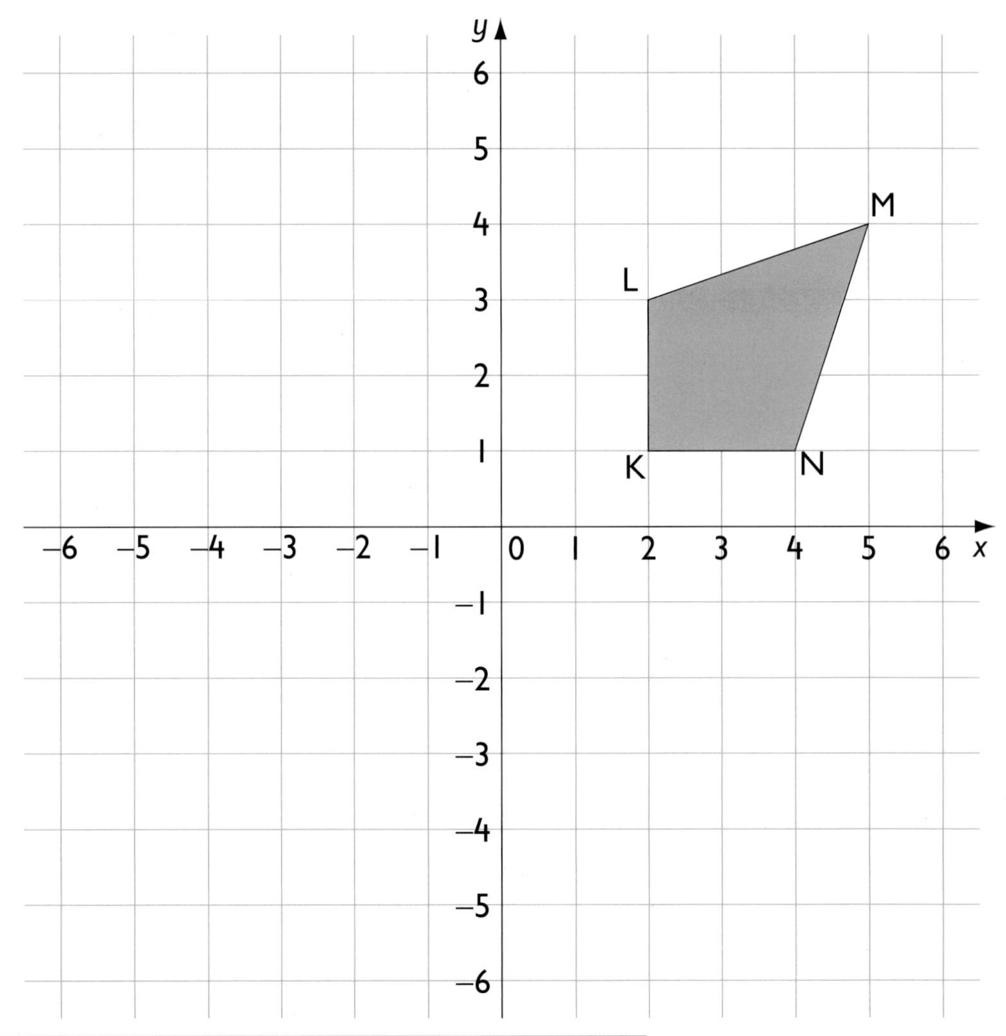

1 Rotate the kite through 90° about the origin (0, 0) into the four quadrants of the grid.

2 Copy and complete the table.

Point	1st quadrant	2nd quadrant	3rd quadrant	4th quadrant
K	(2, 1)	(−1, 2)		
L	(2, 3)			
M				
N				

3 Write what you notice about the co-ordinates of point M.

4 Use your list of 1st quadrant co-ordinates for the kite KLMN.

a Subtract 2 from each x-co-ordinate of kite KLMN.

b Plot the points for the shape PQRS and join them in order.

c Reflect the shape PQRS into the 4 quadrants of the grid.

d Write what you notice about the co-ordinates of point R.

Example
K (2, 1) → P (0, 1)
L (2, 3) → Q (,)
M (,) → R (,)
N (,) → S (,)

5 What if you subtract 2 from each y-co-ordinate of kite KLMN? Investigate.

Measuring and drawing angles

You need:
- a protractor
- a ruler

1 Measure the green and yellow angles.
Calculate the orange angle then measure to check.
Find the angle total for each diagram.

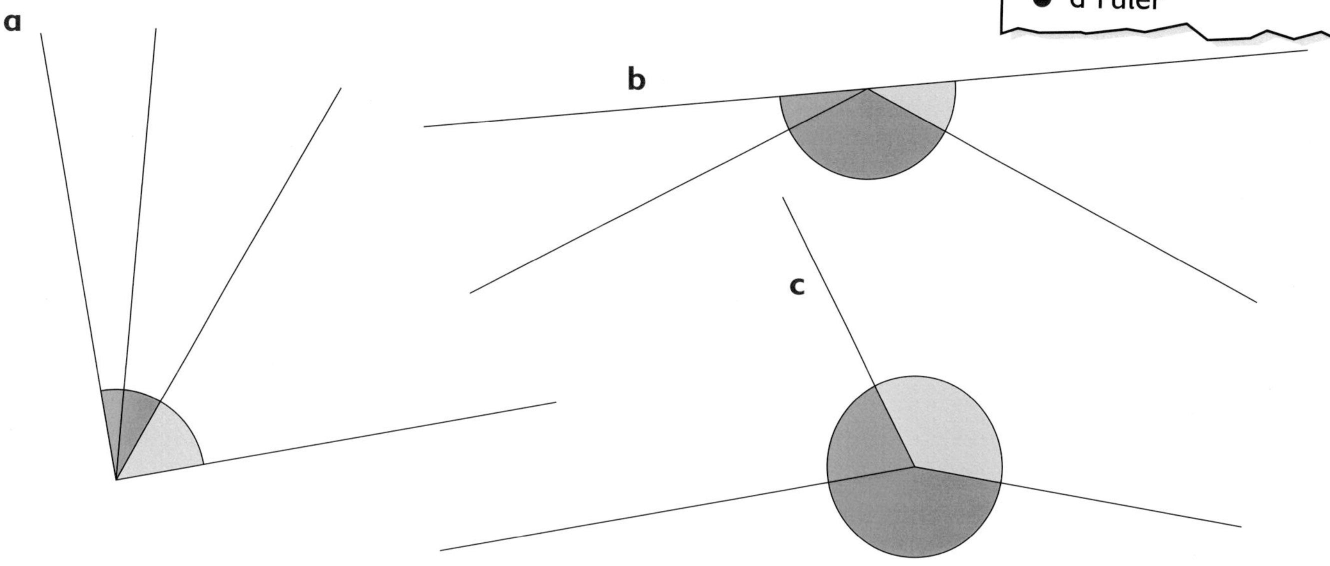

2 Draw and label clockwise and anticlockwise angles of these sizes.

a 65° b 140° c 212° d 298°

3 Make accurate drawings of these angles.

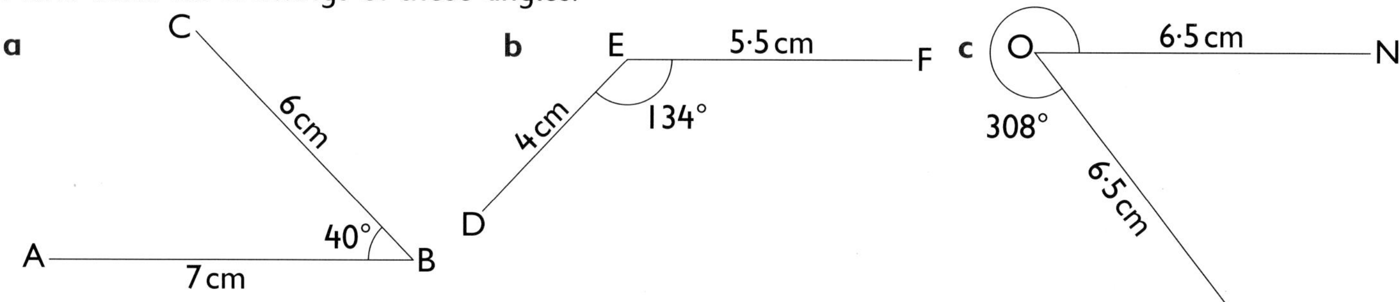

4 Make accurate drawings of these intersecting lines.

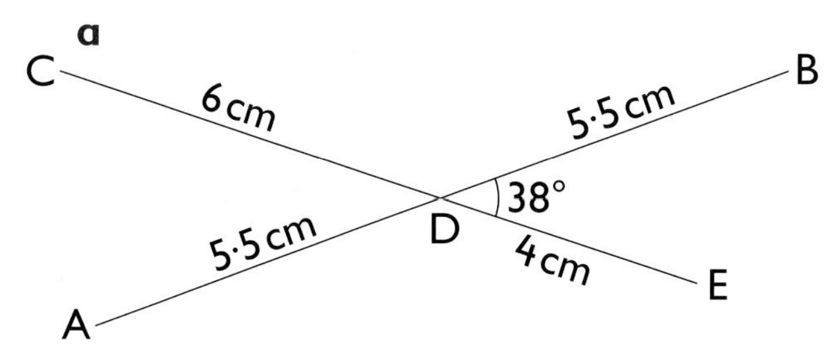

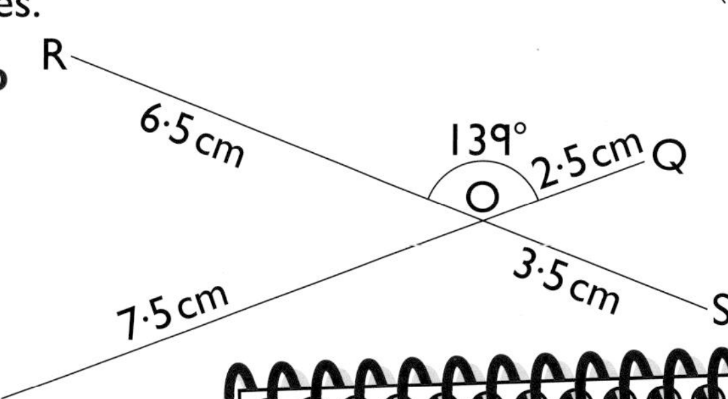

Calculate the size of the other angles.
Measure to check your answers.

5 Work out how to construct a triangular spiral.

You need:
- Resource Copymaster 42
- a set square
- a ruler

Constructing triangles (1)

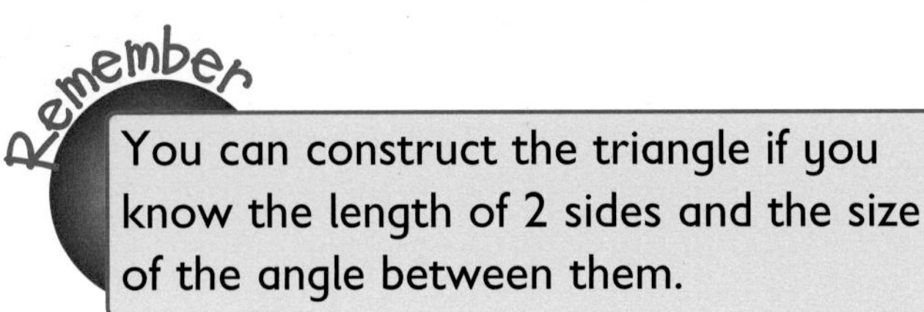

You can construct the triangle if you know the length of 2 sides and the size of the angle between them.

You need:
- a protractor
- a ruler

Step 1

A 7 cm B

Rule a base line AB 7 cm long.

Step 2

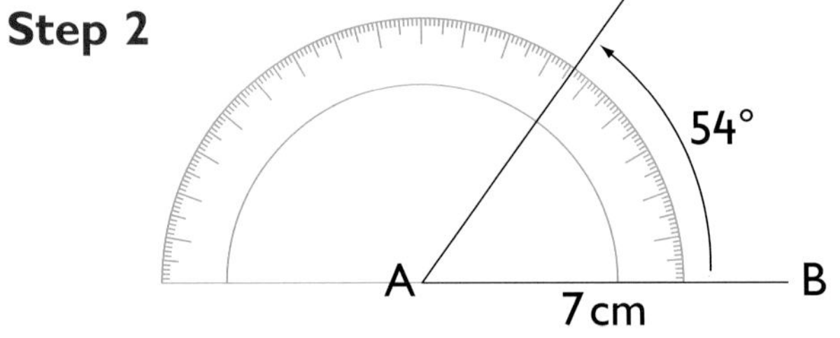

Measuring from the base, mark the position of 54° with a dot. Draw a line at 54° to base.

Step 3

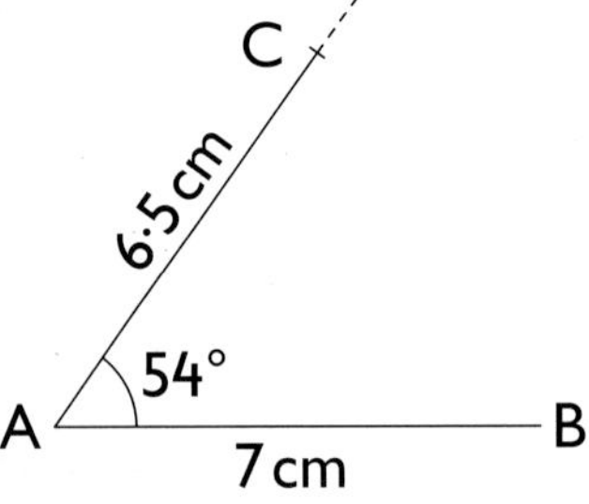

Cut off at 6·5 cm.
Label the point C.

Step 4

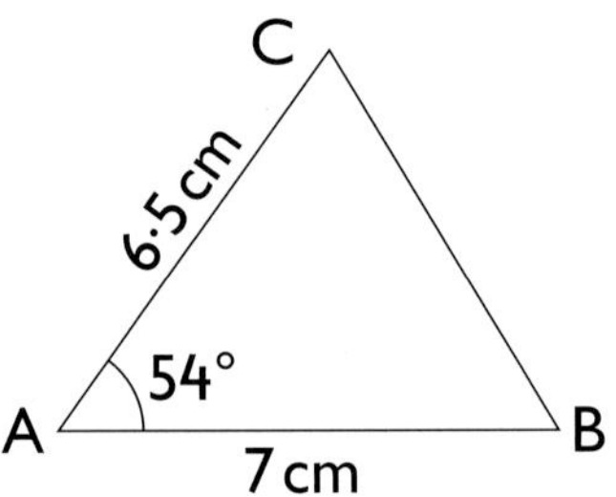

Rule a line to join C to B.

1 Construct $\triangle PQR$ accurately where the base, PQ = 8 cm, PR = 5·5 cm and $\angle RPQ = 65°$.
Label the vertices PQR, the length of the 2 sides and the size of the angle.

Now find the measurements of QR, $\angle PQR$ and $\angle PRQ$.

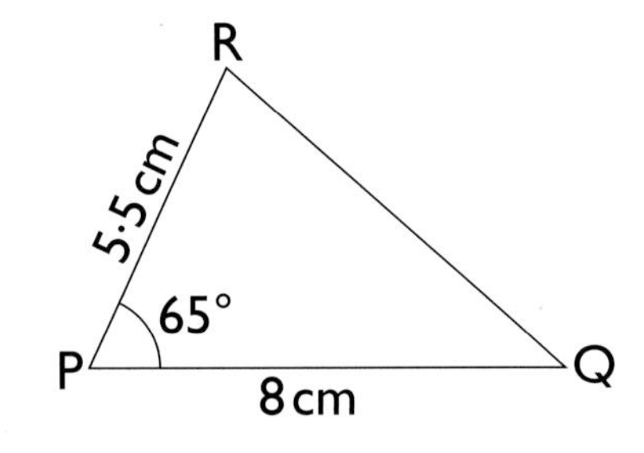

2 Using your ruler and protractor, construct these triangles.

a $\triangle ABC$ where AB = 7 cm, BC = 6.5 cm and $\angle ABC = 45°$.

b $\triangle PQR$ where PQ = 9·5 cm, PR = 5 cm and $\angle RPQ = 72°$.

c $\triangle DEF$ where DE = 6·5 cm, DF = 10 cm and $\angle EDF = 18°$.

d $\triangle KLM$ where KL = 7·5 cm, KM = 6 cm and $\angle MKL = 135°$.

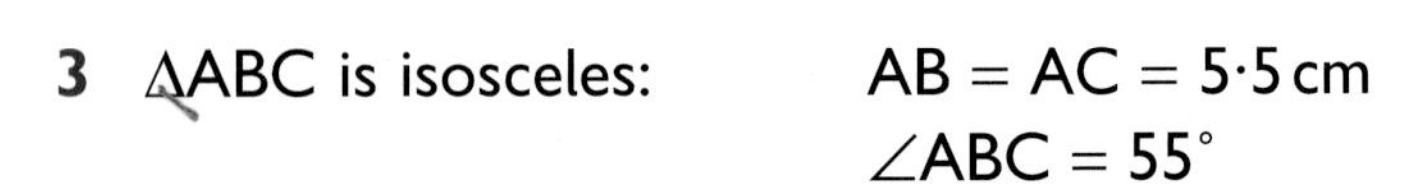

3 $\triangle ABC$ is isosceles: AB = AC = 5·5 cm
$\angle ABC = 55°$

a Find a way to construct the triangle.

b Suppose $\angle BCA = 20°$.
What is the size of $\angle BAC$?

Constructing triangles (2)

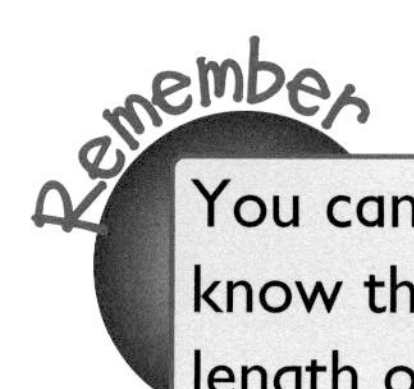

You can construct the triangle if you know the size of 2 angles and the length of the side between them.

You need:

- a protractor
- a ruler

Step 1

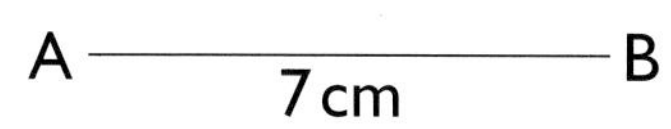

Rule a base line AB = 7 cm.

Step 2

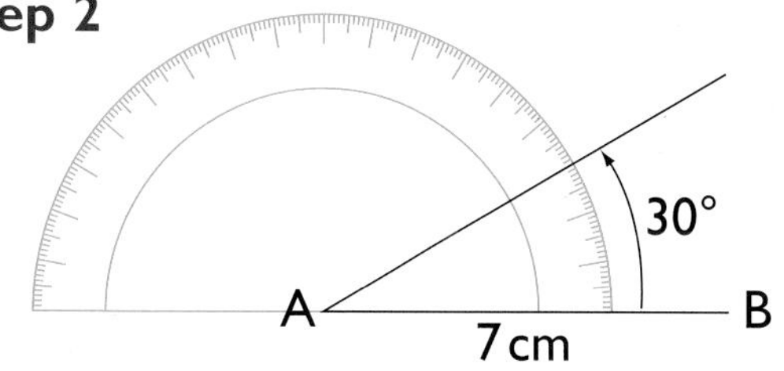

At A, draw a line at an angle of 30° to the base.

Step 3

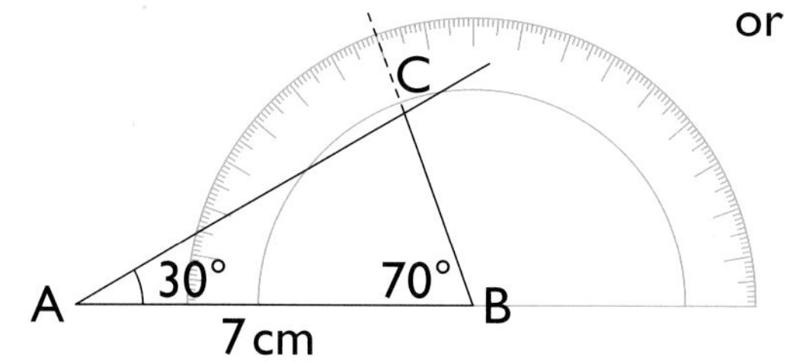

At B, draw a line at 70° to the base.
Extend the sides to meet at C.

or

Step 4

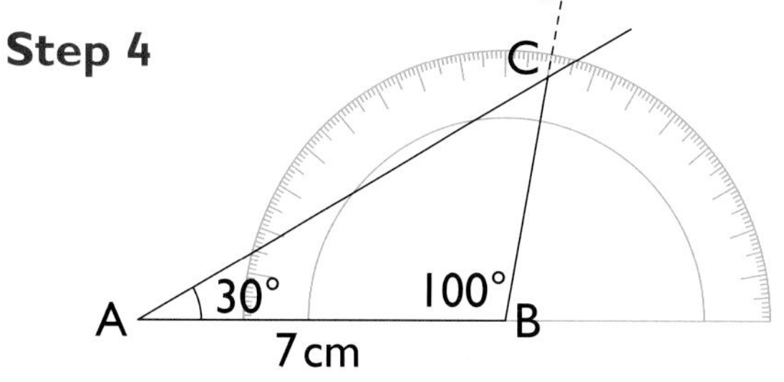

At B, draw a line at 100° to the base.
Extend the sides to meet at C.

1 **a** Construct ΔABC where the base AB = 8 cm, ∠A = 70° and ∠B = 45°.

b Find the length of sides AC and BC.

c Calculate the size of ∠ACB.

2 Use your ruler and protractor to construct these triangles.
Then measure the unknown sides and angle.

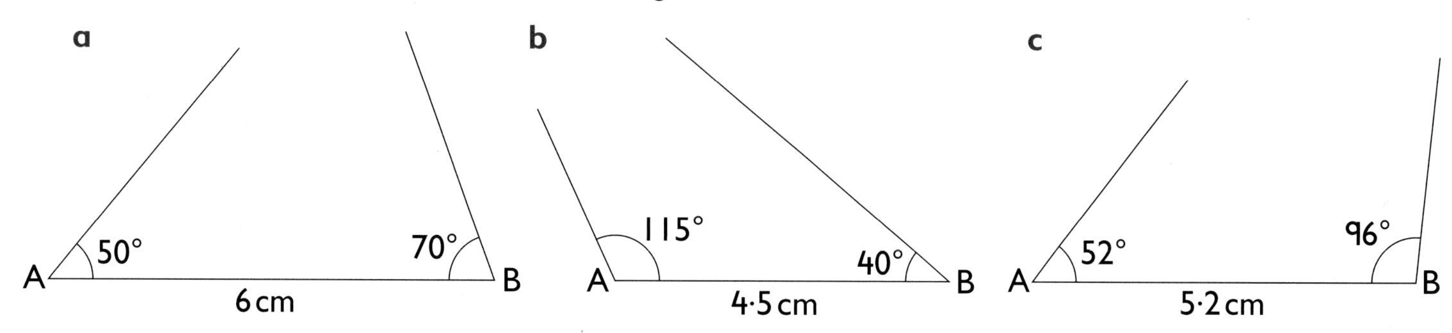

3 **a** ΔPQR is isosceles: One side is 7·5 cm long
The base angles are 65°

Find a way to construct two different triangles.

b Choose 2 angles and 1 side each time.

Target: Make 6 different triangles.

25°	**50°**	**100°**	**7·5 cm**	**5 cm**

Constructing nets of prisms

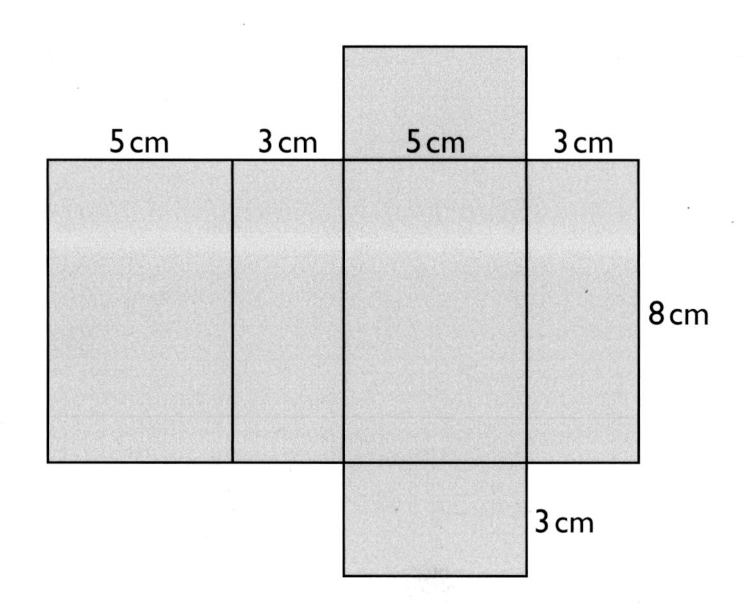

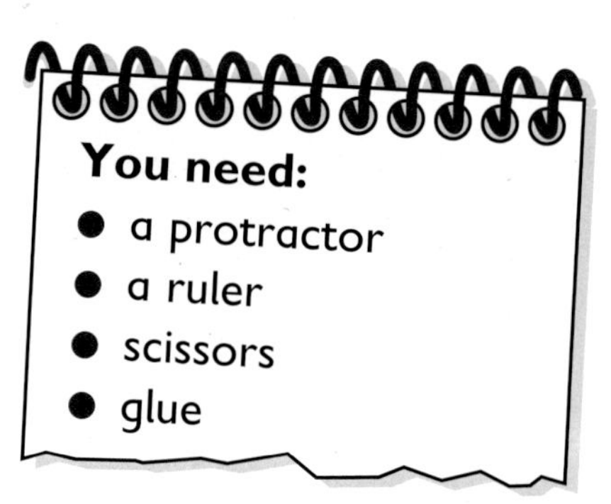

1 Construct on plain paper the net for a cuboid with dimensions 3 cm, 5 cm, 8 cm.
Decide where to draw the 7 tabs.
Score the fold lines and assemble the net.

2 Construct the net for a triangular prism.

- Draw an equilateral triangle with sides of 8 cm.
- On each edge of the triangle construct a rectangle with sides of 8 cm and 3 cm.
- Construct an equilateral triangle with sides of 8 cm on an edge of a rectangle.
- Add the tabs, score the fold lines and assemble the net.

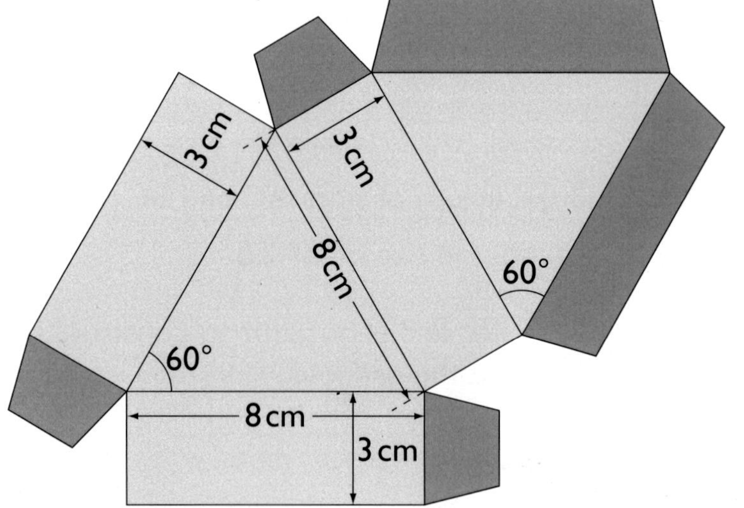

3 a Working with a partner:

- name 2 pairs of parallel sides;
- name 2 pairs of perpendicular sides;
- name the order of rotational symmetry;
- measure ∠EAB and ∠CBA;
- measure ∠FAB and ∠GBA;
- predict and measure ∠F, ∠G, ∠H and ∠I.

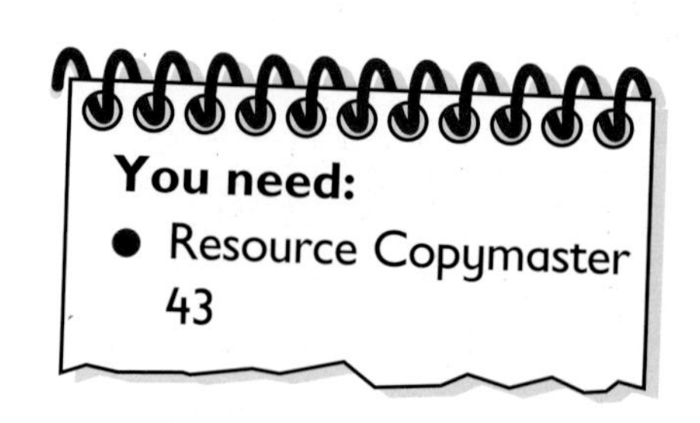

b Carefully cut out your net.

c Score the fold lines and glue the tabs in turn.

d Find a way to place together the two nets to form a cube. Glue together.

Constructing nets of pyramids

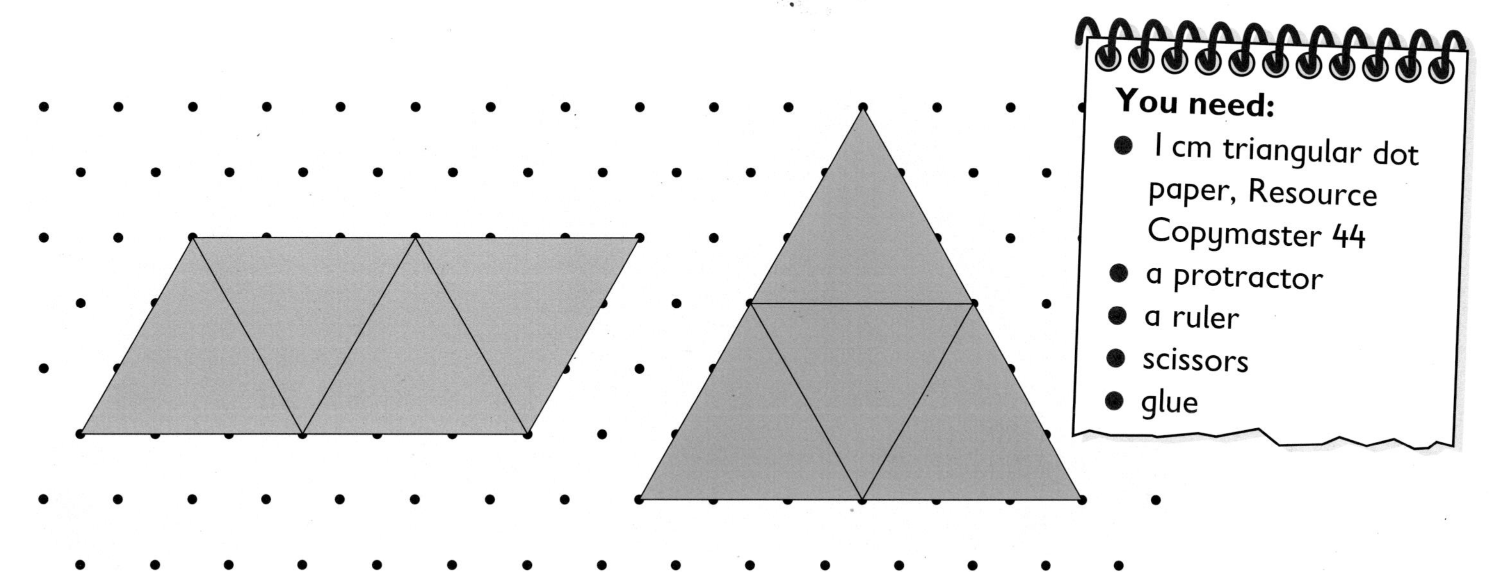

You need:

- 1 cm triangular dot paper, Resource Copymaster 44
- a protractor
- a ruler
- scissors
- glue

1 On 1 cm triangular dot paper, construct both nets for a tetrahedron with sides of 3 cm.
Decide where to draw the 3 tabs.
Score the fold lines and assemble the net.

2 Construct the net for a square-based pyramid.

- Draw a square with sides of 4 cm;
- On one edge of the square, construct an isosceles triangle with a base of 4 cm and angles of 72° then construct a second congruent isosceles triangle;
- On the opposite edge of the square, construct 2 congruent isosceles triangles with bases of 4 cm and angles of 72°;
- Draw the tabs, score the fold lines and assemble the net.

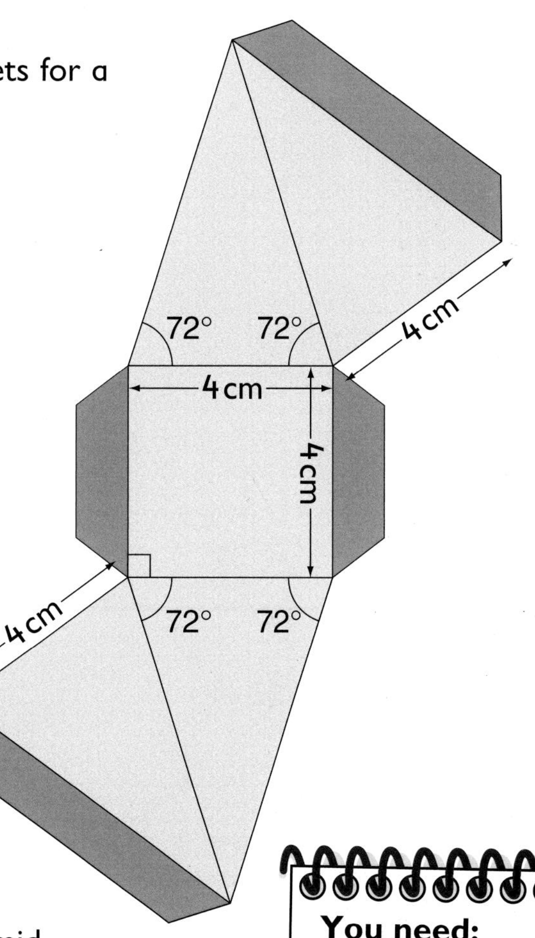

3 Carefully cut out the net on Resource Copymaster 45 for a double hexagonal pyramid.
Assemble the double hexagonal pyramid.
Copy and complete:

You need:

- Resource Copymaster 45

The double hexagonal pyramid has ________ faces, ________ edges and ________ vertices.

● Use names and abbreviations of units of measurement to measure, estimate, calculate and solve problems in everyday contexts involving length; convert one metric unit to another; read and interpret scales on a range of measuring instruments

Lesson 101

Converting units of length

1 The Royal Mint specifies the dimensions of UK coinage.
Measure the diameter of these UK coins to the nearest millimetre.

a

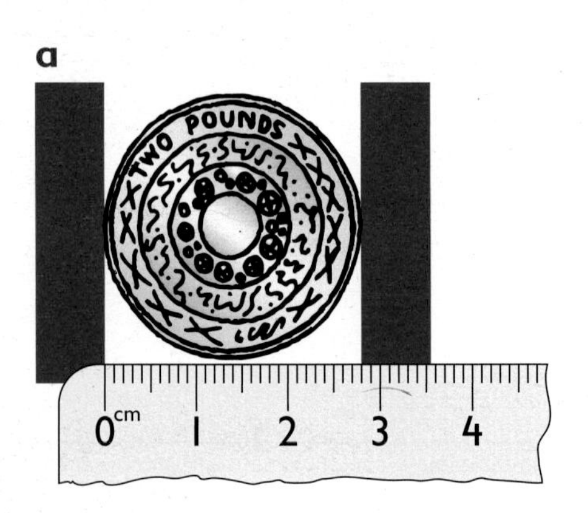

b

c

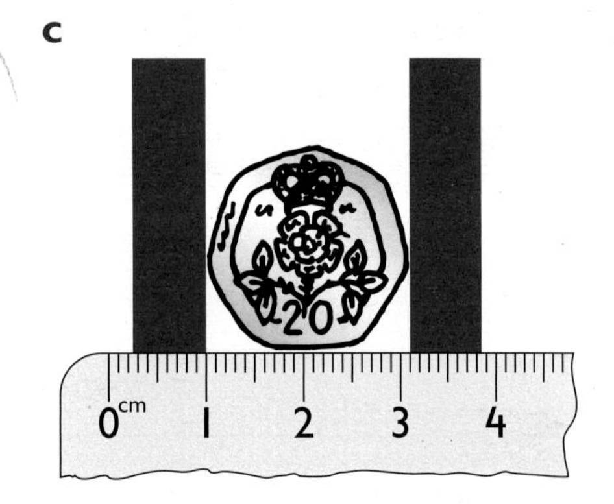

d

e

f

2

Denomination	£2	£1	50p	20p	10p	5p	2p	1p
Edge thickness (mm)	2·5	3·15	1·78	1·7	1·85	1·7	2·03	1·65

Find the height of these stacks of coins.

a £20 in £2 coins

b £20 in £1 coins

c £10 in 50p coins

d £10 in 10p coins

3 Find the length of a line of these coins placed edge to edge in a straight line.

a £10 in 20p coins

b £10 in 5p coins

4 2 schools raise money for charity by placing coins edge to edge in a straight line.

School	Target distance	Diameter of one coin
Lochside	0·1 km in 2p coins	2p 25·9 mm
Glenbrae	0·1 mile in 1p coins	1p 20·3 mm

If both schools reach their target lengths, how much more money does the winning school collect?

- Use names and abbreviations of units of measurement to measure, estimate, calculate and solve problems in everyday contexts involving mass; convert one metric unit to another; read and interpret scales on a range of measuring instruments

Converting units of mass

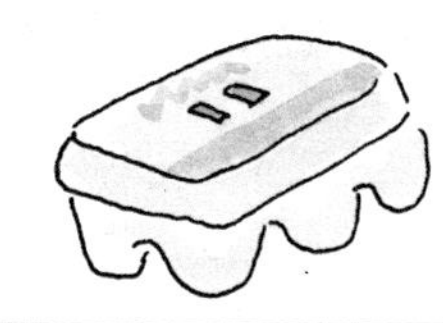

1 At the egg packing station, eggs are weighed, graded and boxed by size.

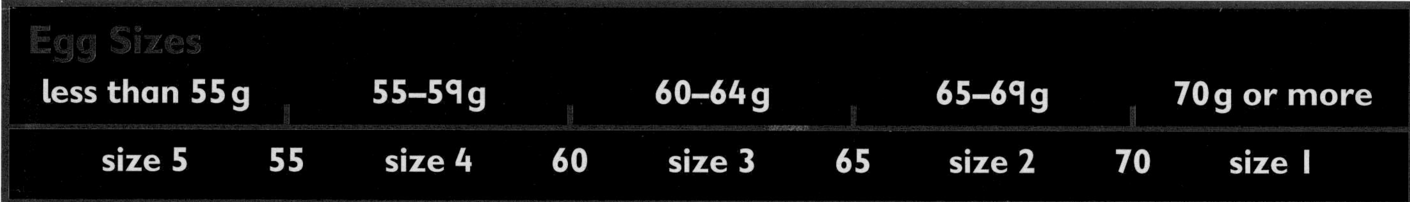

Egg Sizes

less than 55 g	55–59 g	60–64 g	65–69 g	70 g or more
size 5	size 4	size 3	size 2	size 1

a Find the weight and size of these eggs.

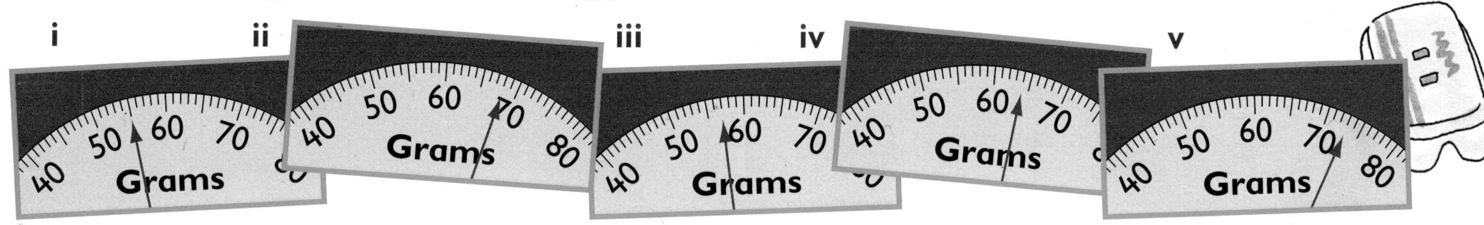

b You buy a box of six size 2 eggs.
Find the range, in grams, that the six eggs might weigh.

c Liz opened a box of 6 eggs and cracked them into an empty bowl.
The empty bowl weighed 125 g.
What size of egg did she use?

2 Three children weighed their backpacks before and after packing their hiking boots. Find the weight of each child's hiking boots:

a to the nearest tenth of a kilogram

b in grams.

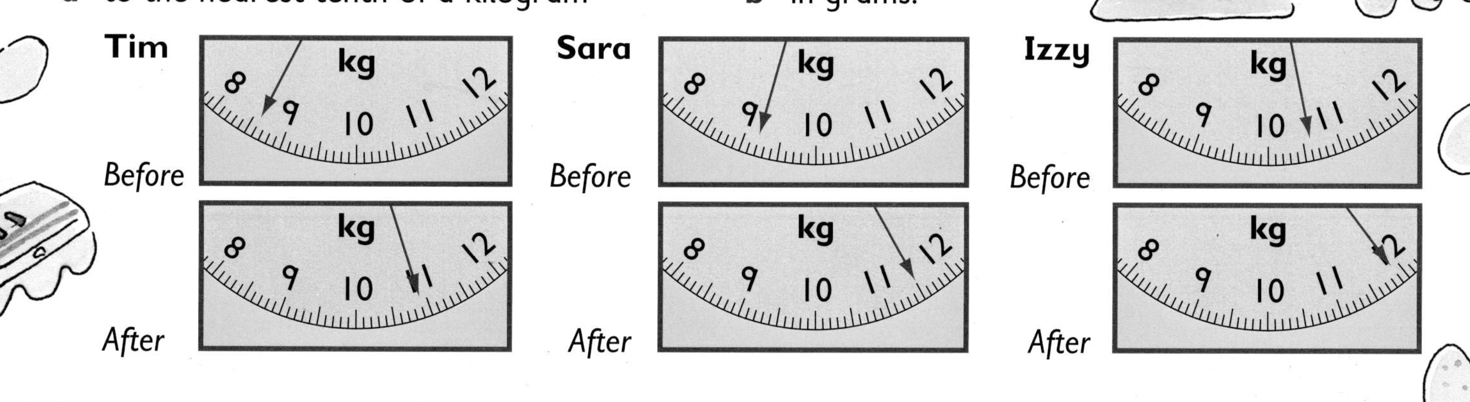

3 Write in pounds the weight shown by each scale.

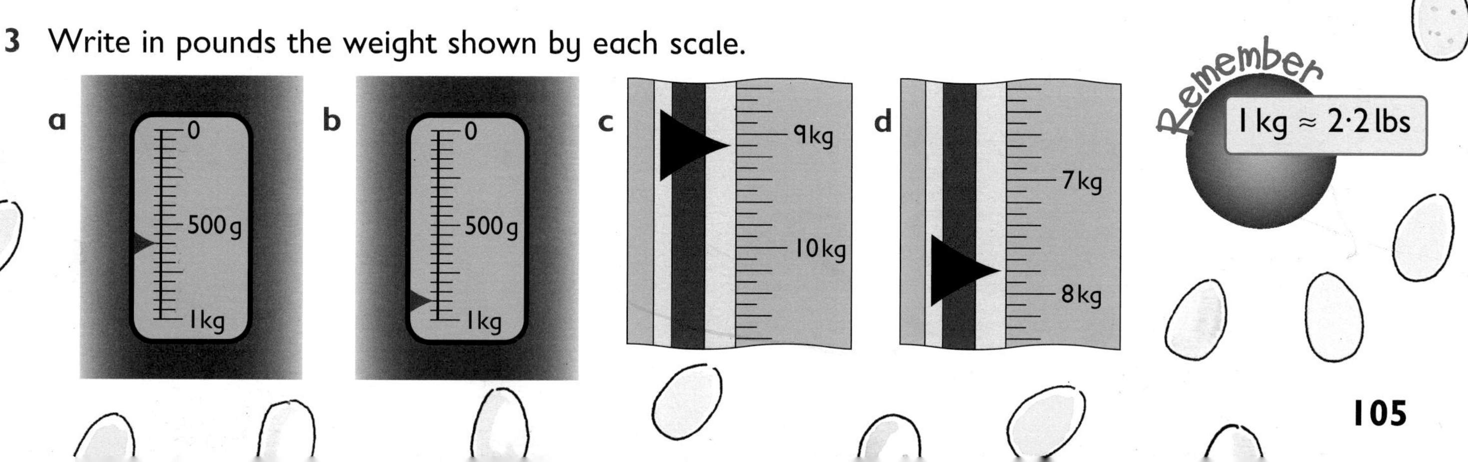

Remember
1 kg ≈ 2·2 lbs

Converting units of capacity

1 Copy and complete the table.
Find the amount of water when the cylinder is filled to these levels.

Level	ml	cl	l
a			
b			
c			
d			
e			
f			

1 l
f
900 ml
e
800 ml
d
700 ml
600 ml
c
500 ml
400 ml
b
300 ml
200 ml
a
100 ml

2 The cylinder is partly filled with water.

a Find how many millitres of water are added to bring the level up to:

i level d **ii** level f

b Find how many centilitres of water are poured off to bring the level down to:

i level b **ii** level a

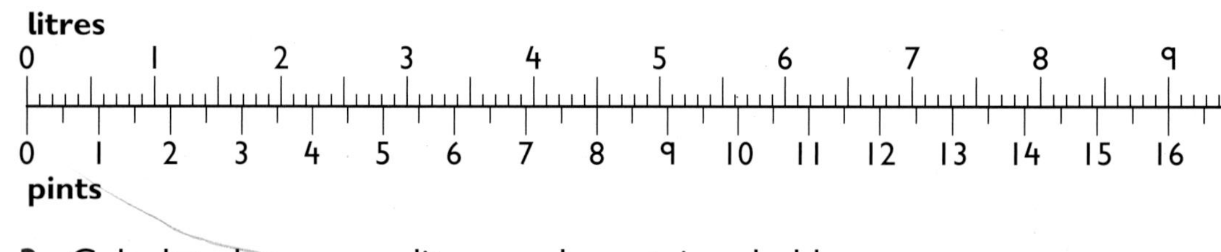

3 Calculate how many litres each container holds.

a

b

c

d

4 Calculate how many litres of petrol each bus uses for a **return** journey to:

a Castle
90 miles

b Airport
108 miles

c Fun Park
175 miles

d Beach
80 miles

World Cup times

1 All World Cup matches were screened live on terrestial television. Using the World time zones map, find the local time for viewers living in the following cities to see the first match. Write the time in two ways.

Example
Cairo 13:30, 1:30 p.m.

- **a** Paris
- **b** New York
- **c** Cape Town
- **d** Sydney
- **e** Rio de Janeiro
- **f** Mexico City
- **g** Singapore
- **h** San Francisco
- **i** Delhi

2 Copy and complete the table for England's group matches.

Game	Time in Tokyo	Time in London
England v Sweden	18:30	
England v Argentina		12:30
England v Nigeria	15:30	

3 There were 64 matches at the World Cup Finals in Korea and Japan.

a A game is 90 minutes long. Excluding any extra time, calculate the total amount of football played altogether:

i in days **ii** in hours **iii** in minutes **iv** in seconds

b A video tape is 3 hours long. A football fan taped all 64 matches. How many tapes did he buy?

Measuring angles

1 **a** Write down the size of:

∠AOB
∠BOC
∠COD
∠DOE
∠FOA

b Name the angle that measures:

100°, 150°, 195°, 260°

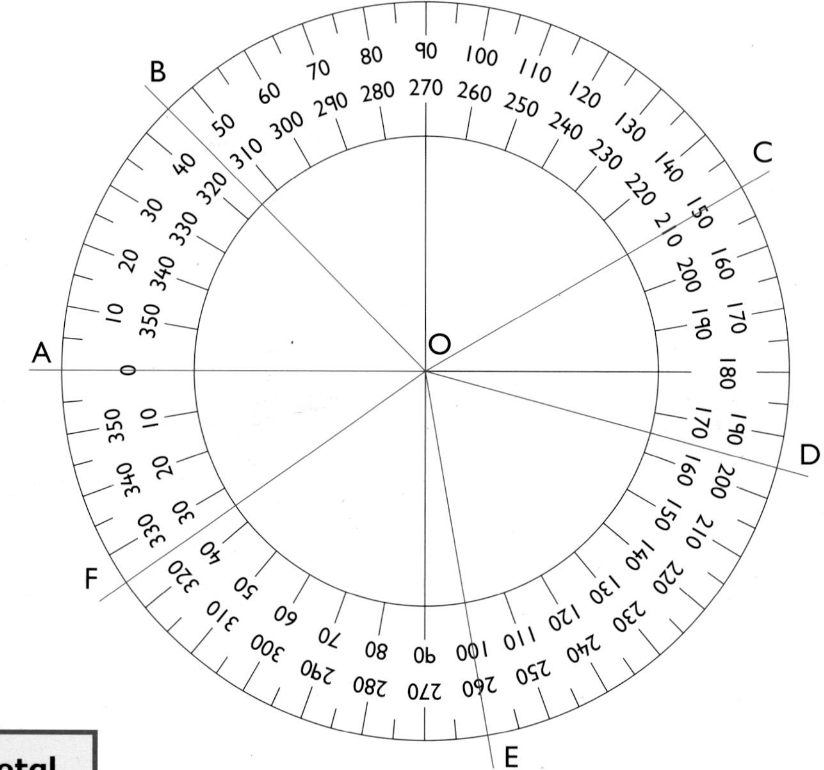

2 Copy and complete the table.

Angle	Obtuse	Reflex	Total
∠AOC	150°	210°	360°
∠BOD			
∠COE			
∠DOF			
∠EOB			
∠FOC			

3 Measure these angles to the nearest degree.

a **b** **c**

4 Play the game, 3 angle pelmanism with a friend.

You need:
- a copy of Resource Copymaster 47
- scissors

Calculating areas and perimeters

1 Find the area and perimeter of these shapes.

2 Work out the area of green for these shapes.

3 These flags are flying at a Visitors' Centre. Calculate the area of white in each flag.

4 Find the area of the red and white stripes in this flag.

Find a way to calculate the total area of the red stripes.

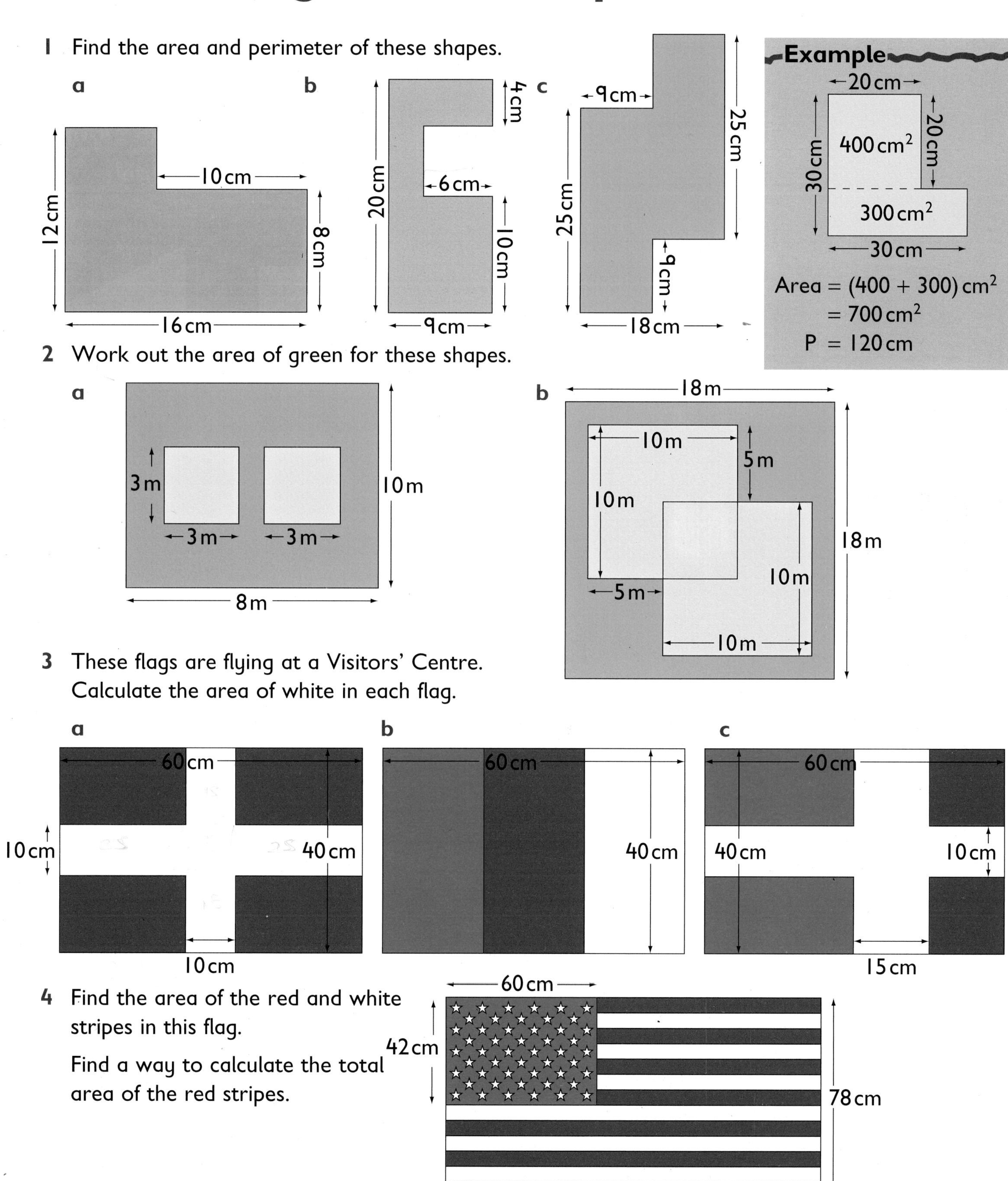

Calculating areas of triangles

You can find the area of a triangle by using the formula:

$$\text{Area} = \frac{1}{2} \times \text{base length} \times \text{height}$$
$$= \frac{1}{2}\text{bh}$$

Example

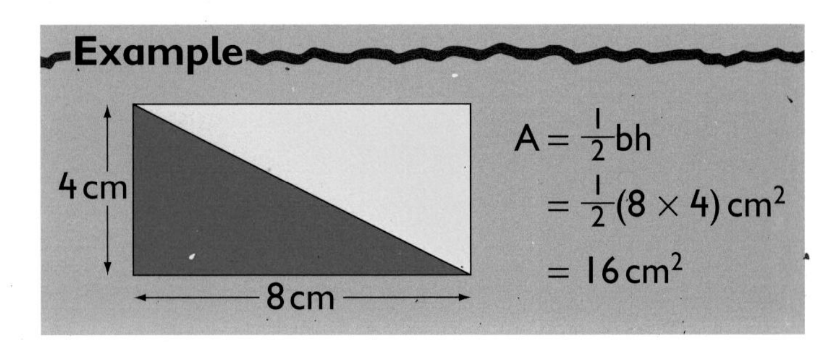

$A = \frac{1}{2}\text{bh}$

$= \frac{1}{2}(8 \times 4)\,\text{cm}^2$

$= 16\,\text{cm}^2$

1 Use the formula to calculate the area of the blue triangles.

a

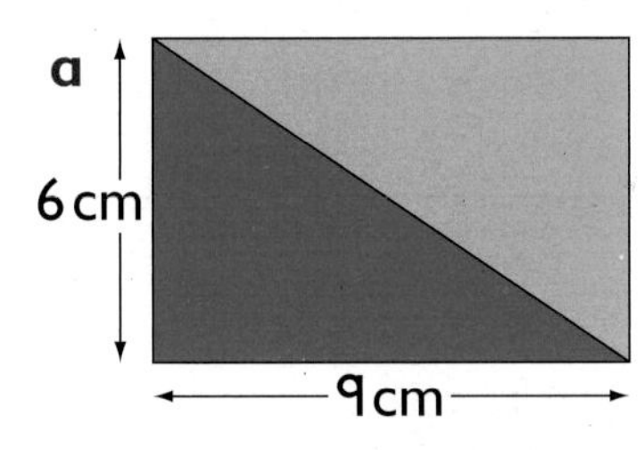

b

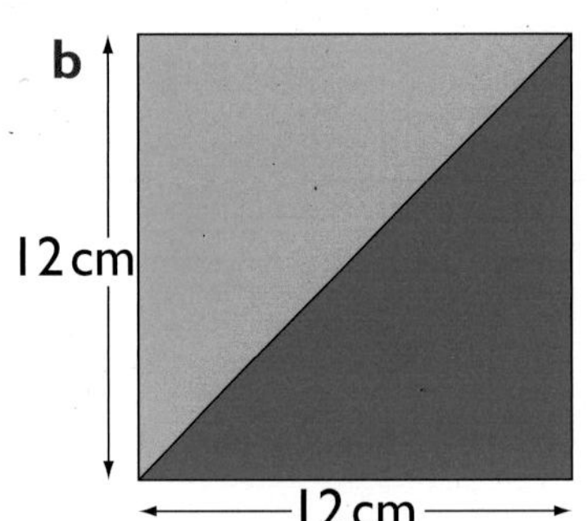

c

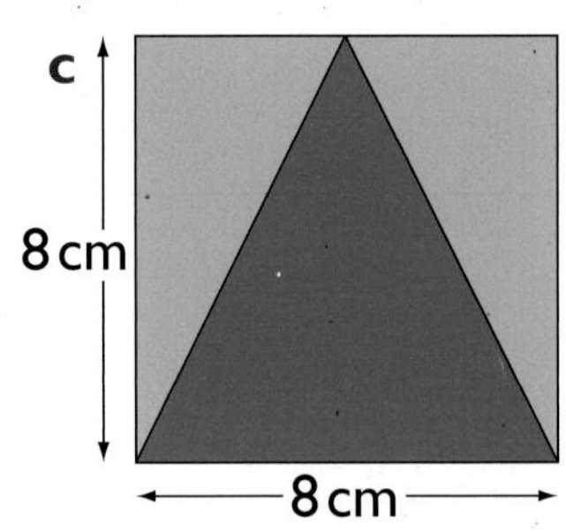

d

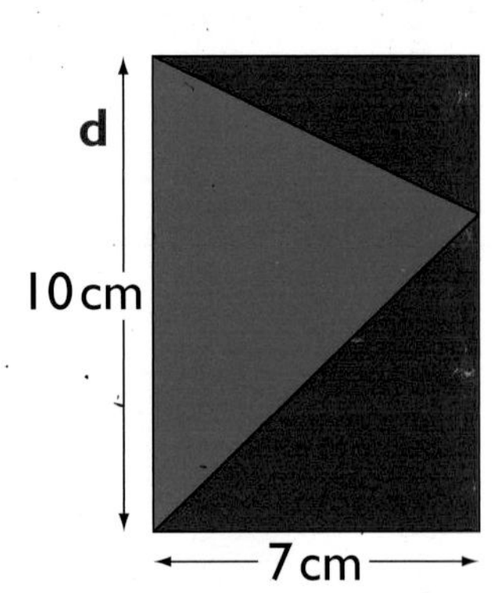

2 You can find the area of the red triangle by subtracting the pieces outside the triangle from the area of the square.

Area of square $= 16\,\text{cm}^2$

Area of pieces $= 8\,\text{cm}^2 + 4\,\text{cm}^2$

$= 12\,\text{cm}^2$

Area of triangle $= (16 - 12)\,\text{cm}^2$

$= 4\,\text{cm}^2$

These represent 1 cm grids.

Calculate the area in cm^2 of the red triangles. Show the steps in your working.

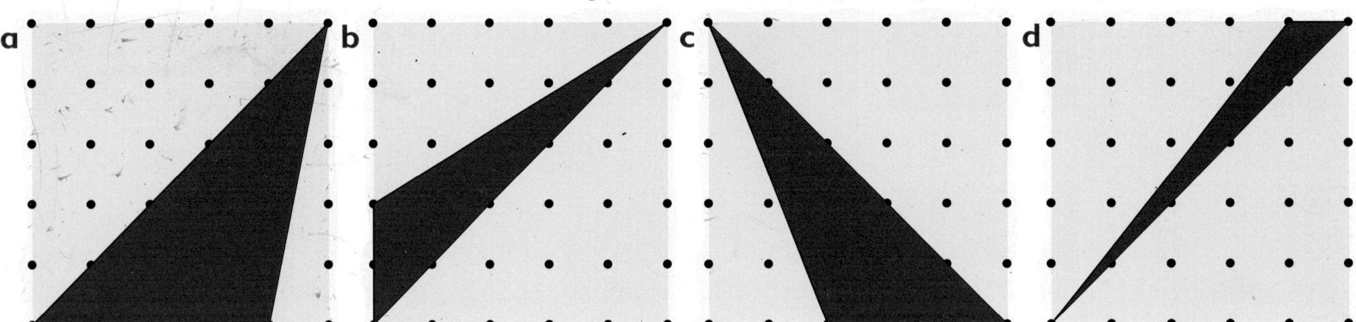

3 Choose methods to find the area in cm^2 of these quadrilaterals.

a

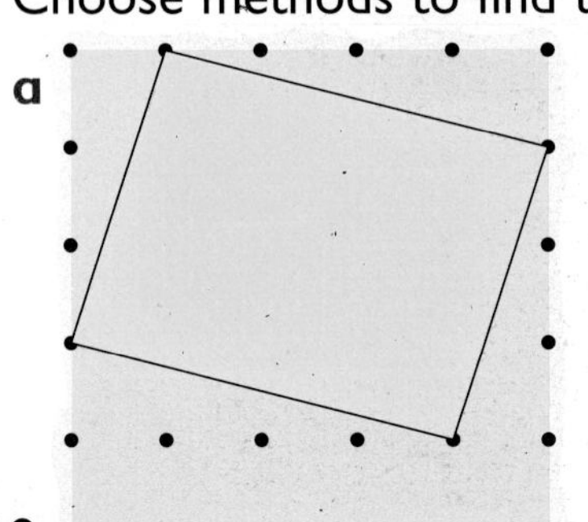

b

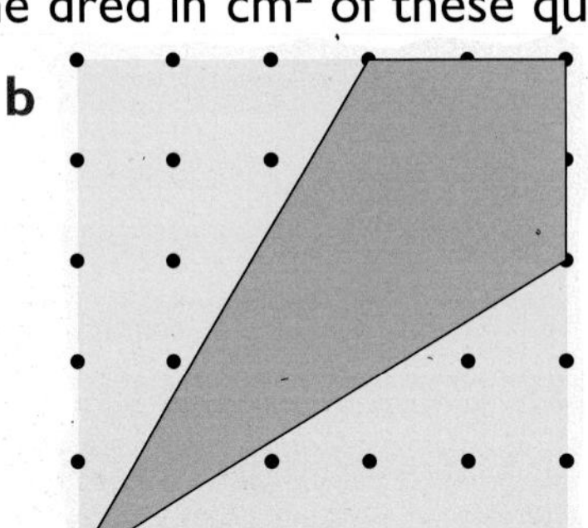

c

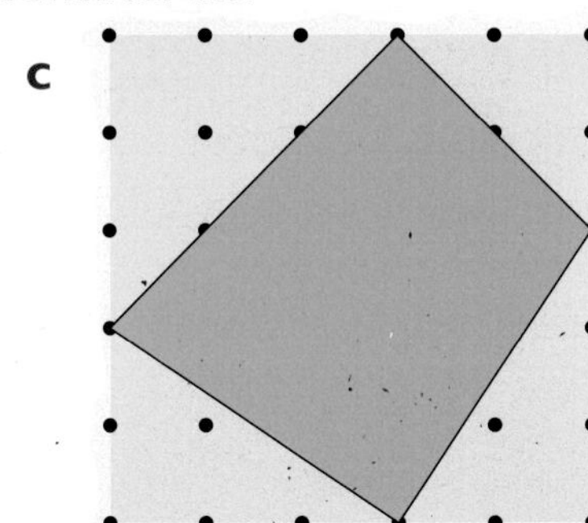

Investigating area and perimeter

1 Mr Hoffman buys 12 identical tiles at the DIY store. He fits them round a square bathroom mirror which has sides of 70 cm.

Calculate the area of one of the rectangular tiles.

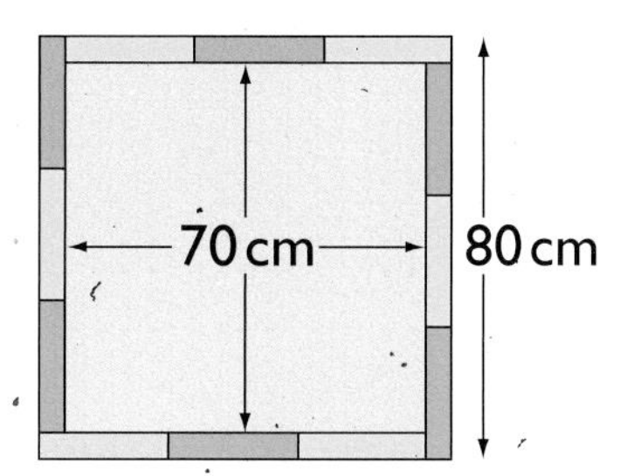

2 This is the floor plan for the entrance hallway of a school.

a Each rectangular area is $12 m^2$. Find the area of the whole floor plan.

b The perimeter of the floor plan is 34 m. Calculate the longest side of each triangle.

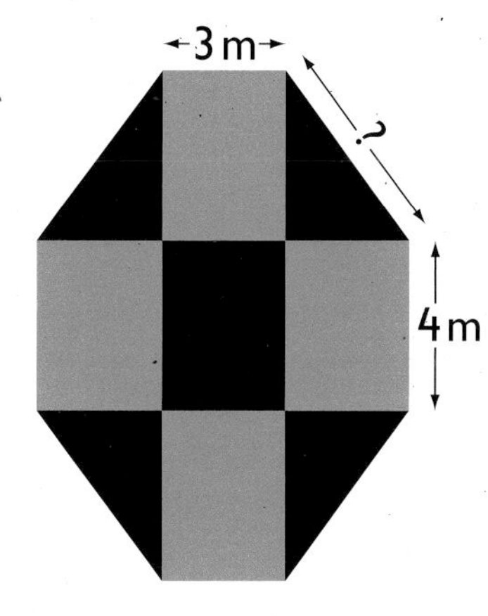

3 Mel made 4 triangles like this. She used them to make a star.

a What is the area of the star?

b What is its perimeter?

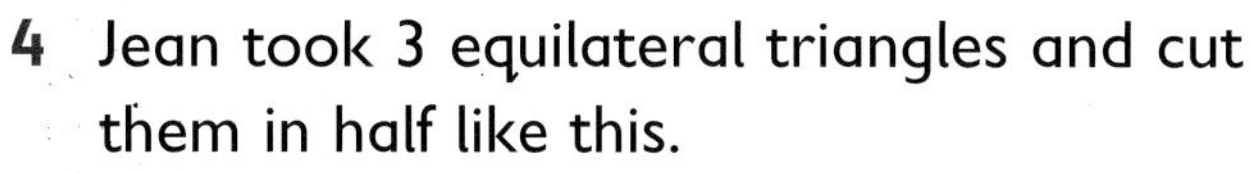

4 Jean took 3 equilateral triangles and cut them in half like this. She used the 6 triangles to make a star.

a What is the area of the star?

b What is its perimeter?

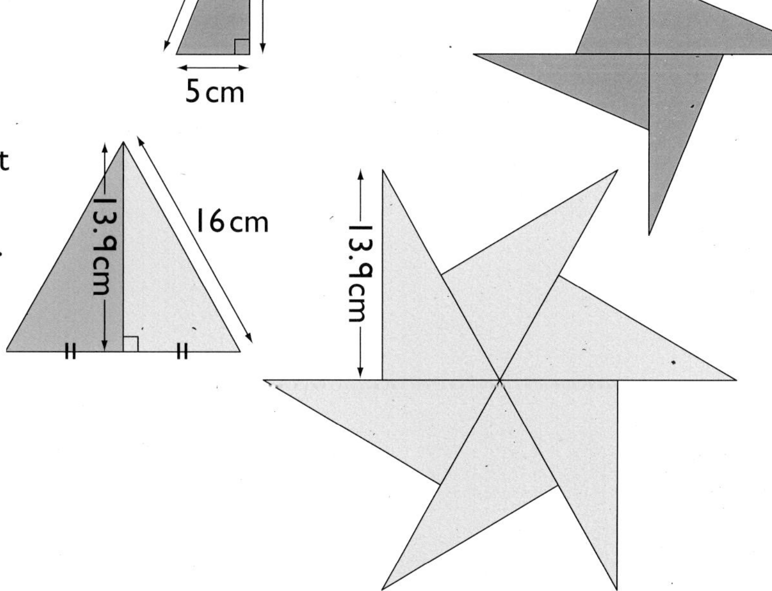

5 This shape has an area of $12 cm^2$ and a perimeter of 20 cm.

a On Resource Copymaster 48, investigate different shapes having a perimeter of 20 cm.

b Find the area of each shape.

c Can you make a shape where $P = 20 cm$ and $A < 9 cm^2$? Explain.

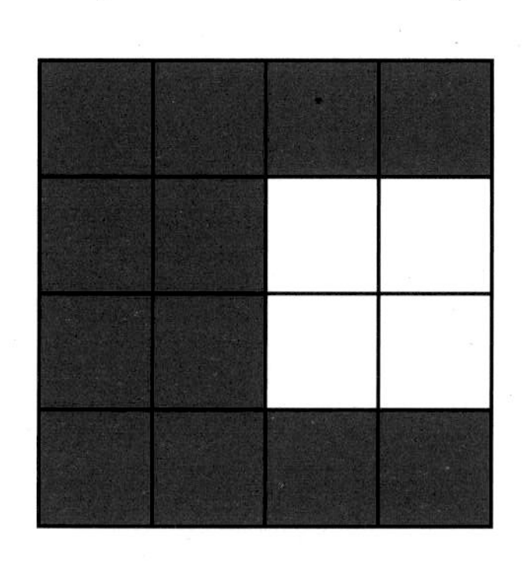

Puzzling areas

You need:
- 1 cm squared paper
- a ruler
- scissors

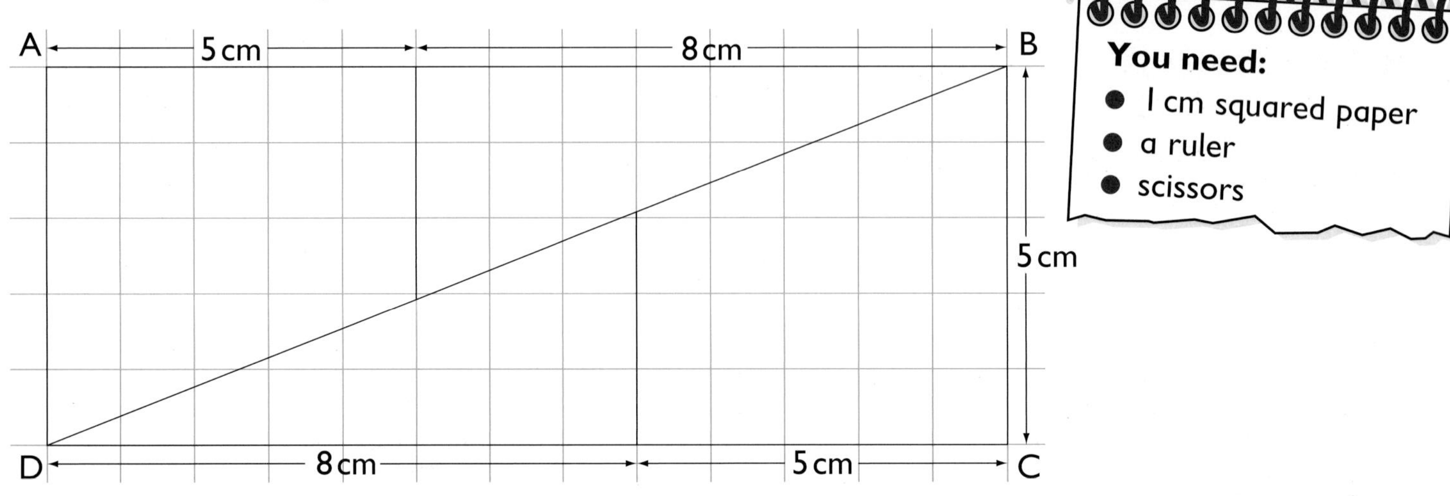

1 Take a sheet of 1 cm squared paper and follow these steps:

- Construct a rectangle 13 cm by 8 cm.
- Draw a diagonal line.
- Count in 5 cm from vertices A and C and drop perpendicular lines.
- Cut out the 4 pieces.
- Rearrange the 4 pieces to form a square 8 cm by 8 cm.
- Find the area of the square and the rectangle.
- Write what you notice.

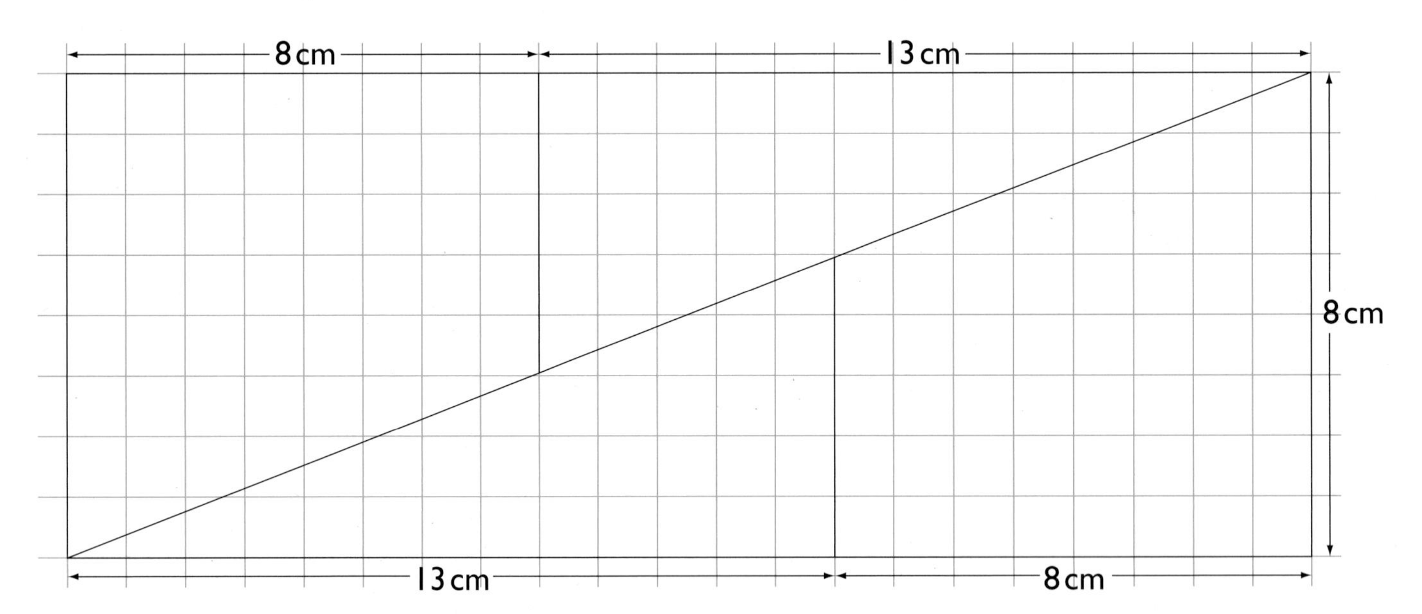

2 Repeat the above steps for a 21 cm by 8 cm rectangle and construct a square with sides of 13 cm.

Find the area of the square and the rectangle. Compare your answers with question 1.

3 What if … the dimensions of the rectangle were 34 cm by 13 cm and the sides of the square were 21 cm? Investigate.

4 The dimensions of the rectangles are based on this sequence of numbers: D = 5, 8, 13, 21, 34, …

What are the dimensions of the next rectangle which will produce similar results?

Calculating surface areas

1 **a** For each cube find:

- the surface area of one face;
- the surface area of the cube.

Copy and complete the table.

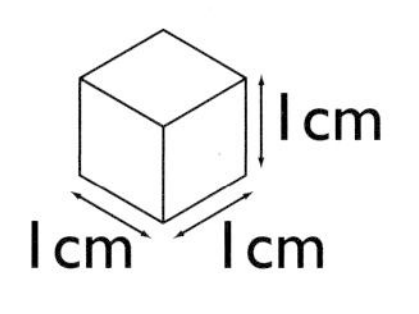

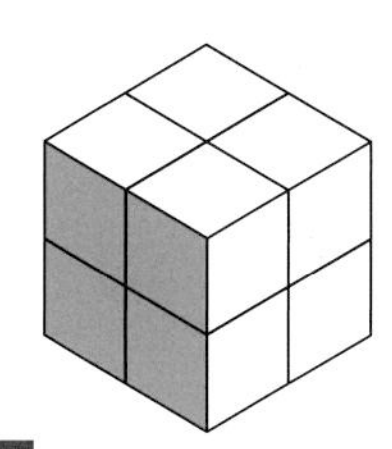

Length of side	1 cm	2 cm	3 cm	4 cm
Area of one face	1 cm^2			
Surface area of cube	cm^2			

b Use the table to find the surface areas of cubes with sides of these lengths:

i 5 cm **ii** 8 cm **iii** 10 cm **iv** *n* cm

2 **a** Calculate the surface area of these cuboids.

A

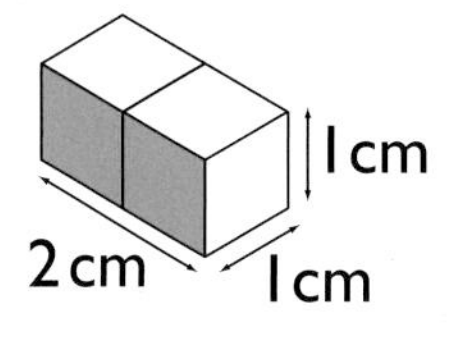

B

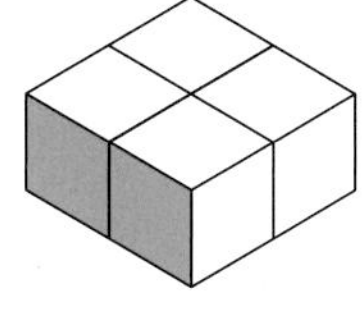

C

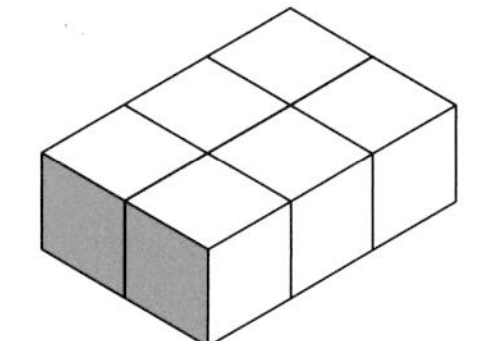

D

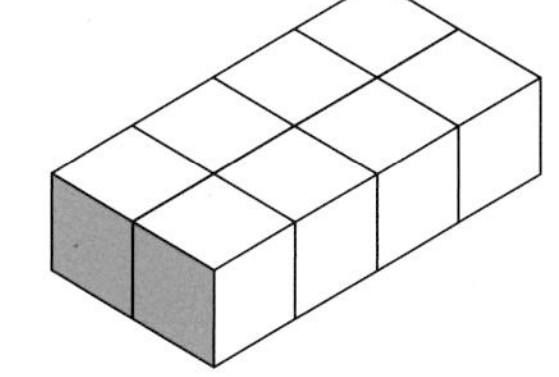

E

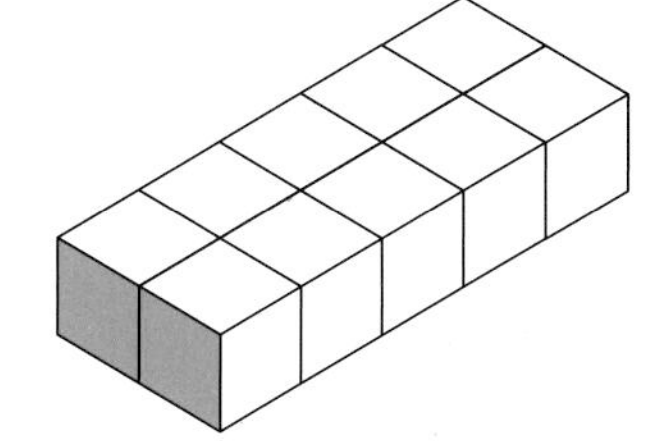

b Find a pattern and use it to work out the surface area of:

i the next 2 cuboids in the sequence;

ii the *nth* cuboid in the sequence.

3 These cubes are made from alternate red and yellow cubes.

a

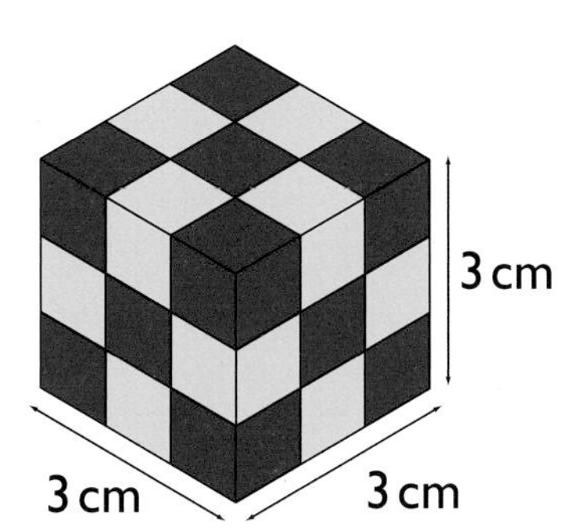

b

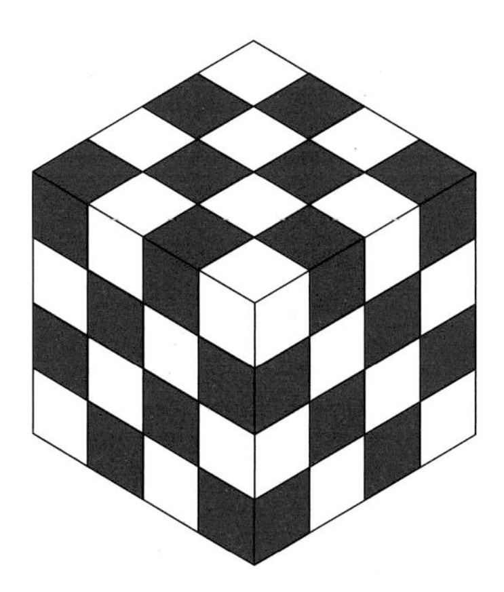

For each cube, find how much of the surface area is red and how much is yellow.

Collecting and organising data

You need:
- the front-page story of a tabloid and a broadsheet newspaper.

1 Work with a partner.
How easy are newspapers to read?

a Decide who will take which paper.

b Begin with the headline.
Count and mark off the first 100 words.

c Record the number of letters in each word in a table.

d Record the data in intervals of 4 letters.

e Record the data from your partner's paper in the table.

	Number of letters (n)				
Name of paper	1–4	5–8	9–12	13–16	17–20
Mail					
Times					

These count as single words:
knock-out *Beckham's*
1998 *20th* *MSP*

2 **a** Record the same data again in intervals of 3 letters.

b Compare your results with question 2.
Which are more useful: intervals of 3 letters or intervals of 4 letters? Why?

3 Choose one of these topics. List the data you would need and the units of measurement you would use.

1 *Amount of water wasted in one day by a dripping tap.*

2 *Number of words in a dictionary.*

3 *Strength of handles of plastic bags from supermarkets.*

Collecting data from experiments

Practical activities for 2–3 children

You need:
- 10 centimetre cubes

1 *Which is your better catching hand?*

   a Experiment

   - Place 10 centimetre cubes on the back of your writing hand.
   - Toss them gently upwards, turn your hand round quickly and catch as many as you can.
   - Repeat for the non-writing hand.

   b Record the results in a frequency table.

Cubes caught (c)	0	1	2	3	4	5	6	7	8	9	10
Writing hand											
Non-writing hand											

   c Repeat the experiment another 9 times.

   d Compare your results with your partner's.

2 Choose one of the following questions to investigate.

Do taller people take a larger shoe size?

Do taller people have a longer arm span?

Can taller people hold their breath longer than shorter people?

- Decide what data you need for your survey.
- Design a frequency table.
- Record the data in the table.

You need:
- tape measure
- metre stick
- seconds timer or stopwatch

3 In the game of Scrabble, 42 out of the 100 tiles are vowels.

   a Experiment

   - Choose a paragraph from a book.
   - Count off the first 100 words.
   - For each word, record the individual vowels.

   b Compare the frequency of the vowels with their frequency in Scrabble.

   c Repeat the experiment with a different source, for example, newspaper, magazine.

Scrabble tiles	
Vowel	Frequency
A	9
E	12
I	9
O	8
U	4

You need:
- a book

Mode and range

Example

For 1, 2, 3, 3, 4, 6, 9 the mode is 3
For 4, 2, 2, 3, 4, 6 the modes are 2 and 4
For 7, 2, 5, 5, 7, 2 there is no mode

1 Find the mode of each set of cards.

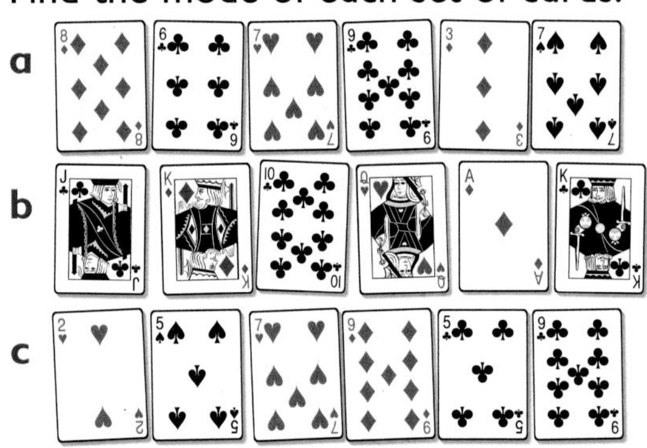

a

b

c

d

The mode is the most common value.

The range is largest value minus smallest value.

2 Arrange the data in order from smallest to largest. Then write down the mode and the range.

a

b

c

d

3 a Find the range and modal height.

b Find the range and modal age.

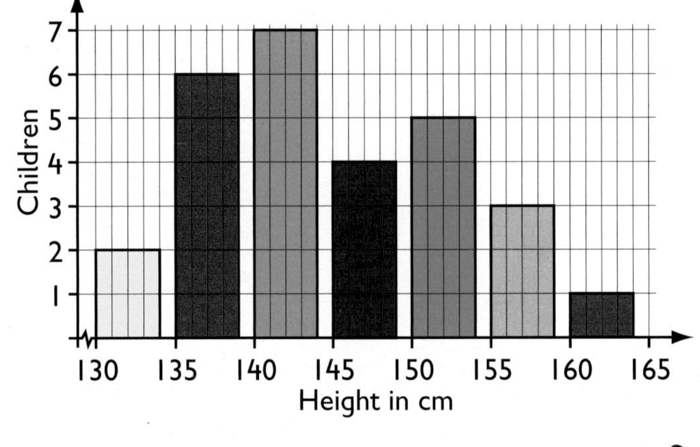

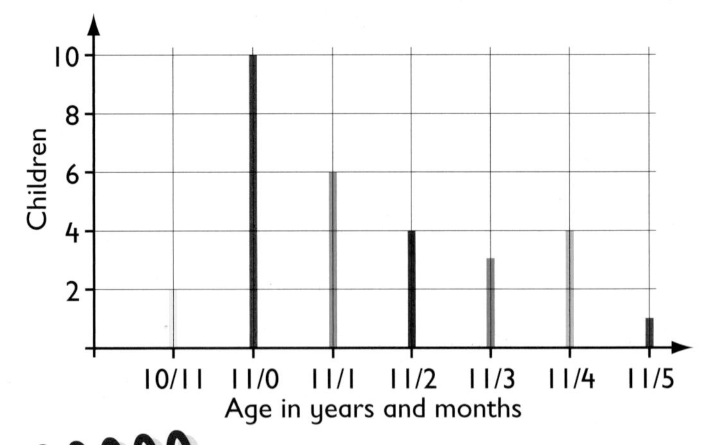

4 Tim is a keen skier. The table gives the weather record for his favourite resort in Colorado.

a Find the range and modal temperature.

b Draw a graph to check your results.

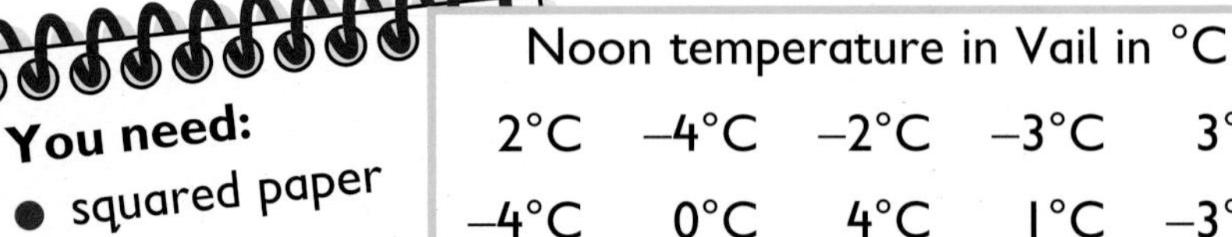

Noon temperature in Vail in °C

2°C	–4°C	–2°C	–3°C	3°C
–4°C	0°C	4°C	1°C	–3°C
4°C	–2°C	3°C	–1°C	2°C
3°C	–1°C	0°C	3°C	–2°C

Median and range

1 Write these scores in order, beginning with the smallest.
Then find the median value for each set.

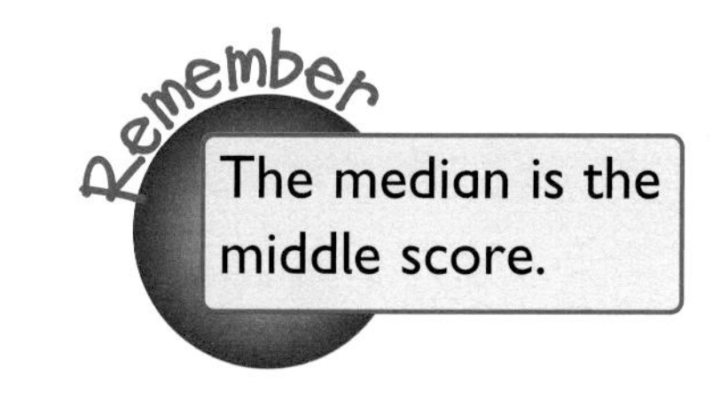

Amy 7, 5, 9, 4, 8, 6, 7
Bert 4, 8, 5, 7, 5, 6, 8
Chris 5·7, 6·3, 7·2, 6·0, 5·9, 6·8, 6·3
Diane 62%, 57%, 71%, 68%, 75%, 64%, 69%

2 Find the median of these values

a 107, 106, 106, 107, 105, 107, 108, 106

b £90, £70, £80, £60, £80, £60, £70, £70

c 0·8, 0·5, 0·8, 0·5, 0·6, 0·8, 0·6, 0·8

d 7·5 g, 6·8 g, 7·4 g, 6·9 g, 7·3 g, 7 g, 7·2 g, 7 g

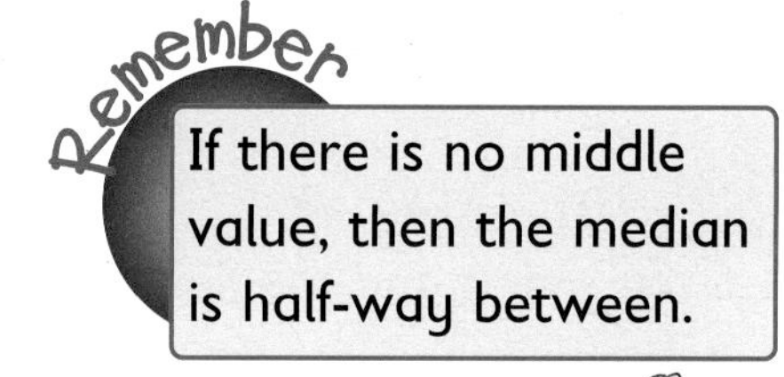

3

Name of duck	Length (cm)	Size of clutch	Weight: egg (g)	Days to hatch
Coot	38	6–9	27	24
Gadwall	51	8–12	43	26
Mallard	58	9–13	50	28
Pintail	70	7–11	47	23
Shelduck	62	8–12	78	29
Shoveller	48	8–12	39	23
Teal	38	5–6	28	22
Wigeon	49	7–10	47	23

a Find the median for:
- length of duck
- weight of egg
- days to hatch

b Find the mode for:
- length of duck
- weight of egg

c Find the range for:
- length of duck
- size of clutch
- egg weight
- days to hatch

4 Design a table to show the facts about these ponies.

Jenny

Height
median 13 hands
mode 12 hands
range 3 hands

Julie
Jingle

Age
median 5 yr 6 mth
mode 6 yr 2 mth
range 18 months

Joyce

Prizes won
median 23
mode 25
range 10

Mainly means

Remember To find the mean, add the values and divide by the number of values.

1 Calculate the mean of these numbers.

a 11, 13, 15
b 81, 92, 103
c 46, 42, 48, 44
d £9, £19, £29, £39
e 2·7, 4·6, 7·5, 9·2, 7·0
f 17%, 26%, 35%, 44%, 52%

2 Find the mean price of these purchases.

3 Find the mean and median weight of these parcels.

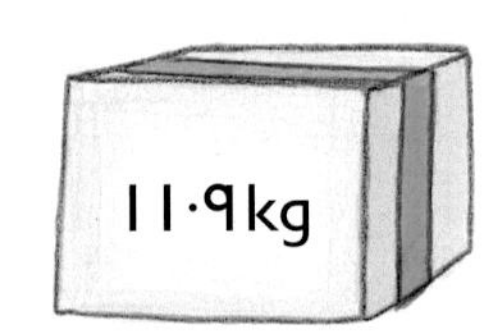

4 Calculate the mean and range of liquids in these containers.

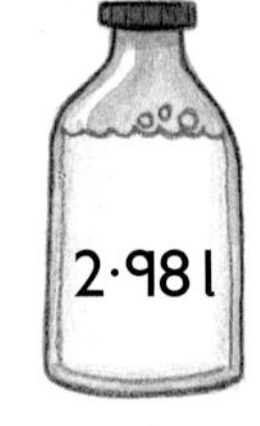

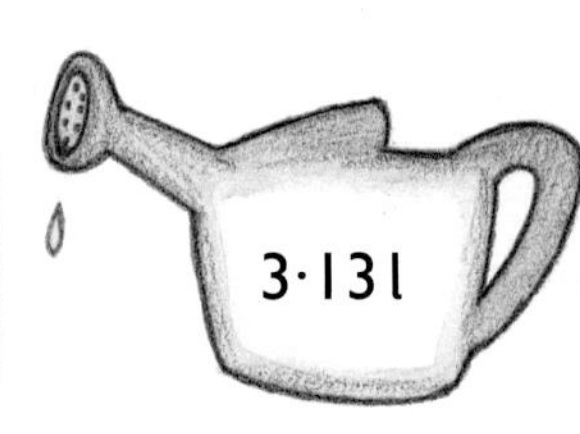

5 The table shows the distances covered by each athlete in 5 days of training.

a Calculate the mean, median, range and modal distance.

b Calculate the mean distance covered by each athlete in one day's training.

Athlete	Distance in 5 days
Kevin	33·1 km
Alun	28·7 km
Meena	23·8 km
Penny	23·8 km
Kapil	32·9 km
Deng	30·5 km

Buy our **Everlong P9** *battery.*
Average life of 30 hours – Guaranteed.

6 a Find the mean of this sample of P9 batteries.

b Make a frequency table.

c Decide whether the claim is true or false.

P9 Battery life in hours

29	31	30	28	27
26	30	32	29	31
28	26	29	30	32
30	31	28	33	30

Which holiday?

1 The Parkers and the Browns are neighbours.

Last summer they took their holidays in the same week in August.

The Parkers went to Palmy Beach and the Browns to Breezy Point.

a Which family had more hours of sunshine per day? Calculate the range and mean number of hours of sunshine for one day.

b Write two sentences comparing the results. Use the range and the mean.

Hours of sunshine

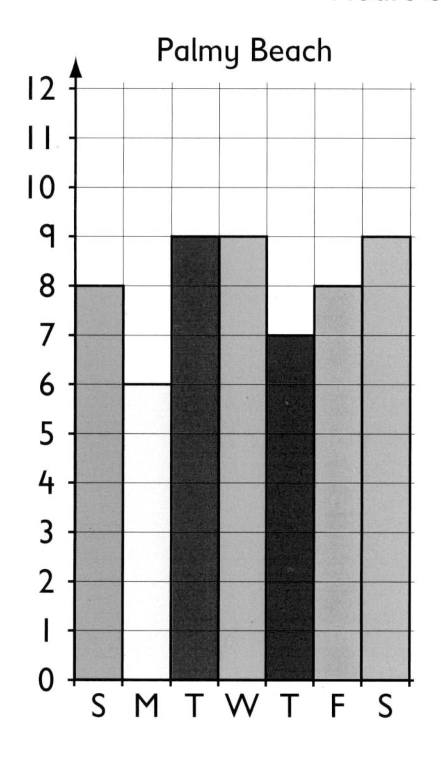

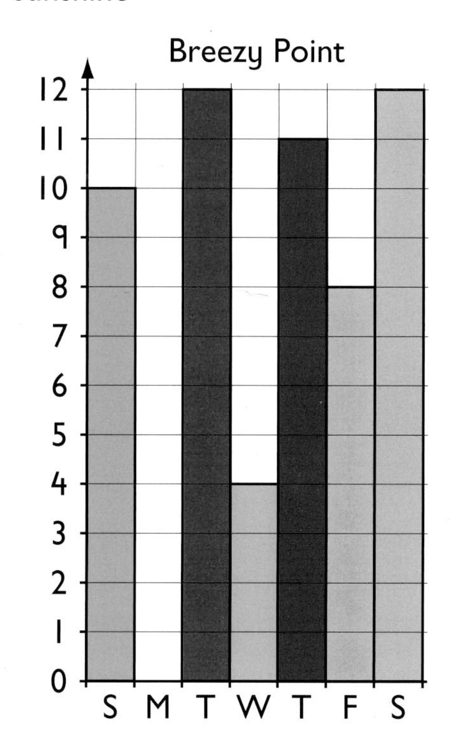

2 **a** Calculate the range, mean and modal temperature for each holiday resort.

b Compare the results.
Use the range and the mode.

Temperature at noon °F

Resort	Su	M	Tu	W	Th	F	Sa
Palmy Beach	77	72	78	77	75	79	76
Breezy Point	75	82	81	78	81	80	81

3 Decide which holiday you would prefer and why.

4 The Waddells went on a cruise. The ship posted the daily noon temperatures.

Gillian kept a record for one week and worked out these statistics for 'Silverglide'.

Work out the temperature for each of the 7 days.

Our Cruise

Temperatures at noon °F

Mean = 78°F Range = 8°F

Mode = 80°F Median = 78°F

The highest temperature was 82°F.

- Construct, on paper and using ICT, graphs and diagrams to represent data …
- Interpret diagrams and graphs …
- Write a short report of a statistical enquiry and illustrate with appropriate diagrams, graphs and charts …

Constructing and interpreting bar charts

1 Which computer screen at Bert's the Chemist shows the monthly sales of:

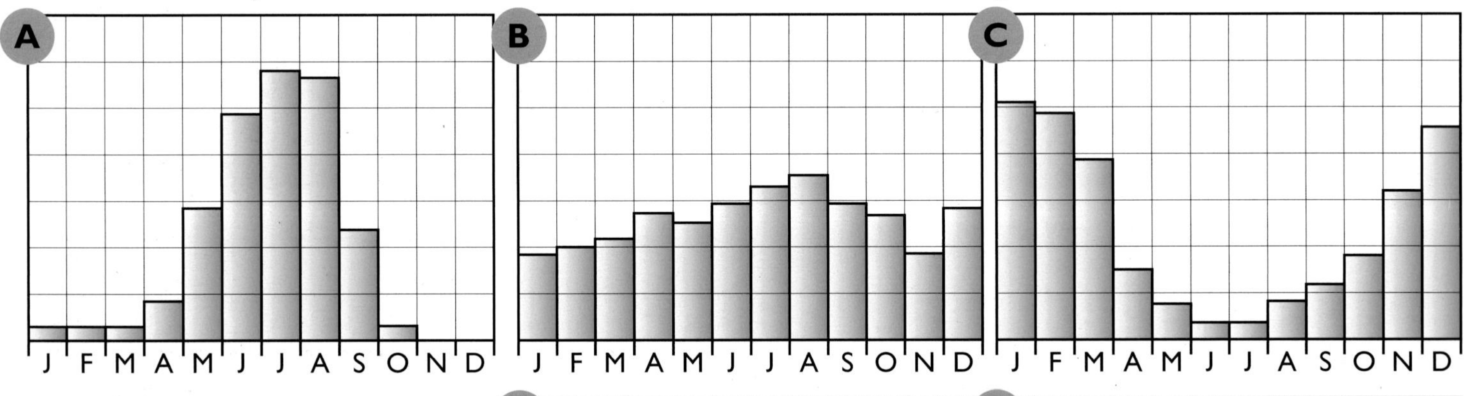

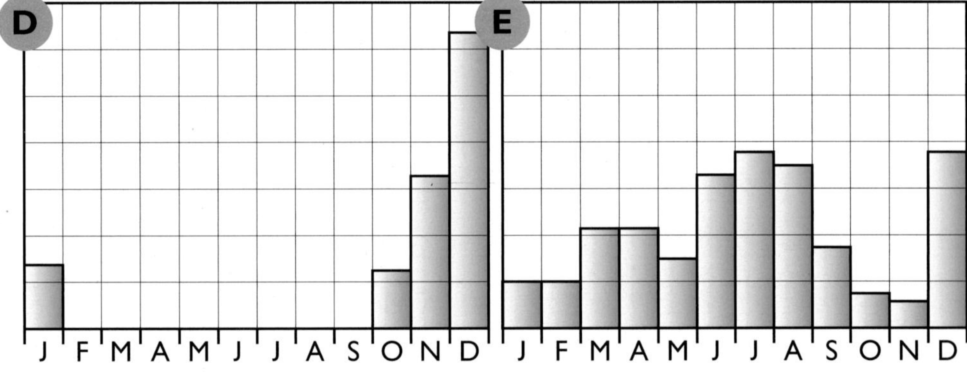

a shampoo?

b cold remedies?

c sun lotion?

d film?

e Christmas gifts?

Justify your answer.

2 200 children in Year 6 took part in a survey. They were asked: *How long did you spend going on line after school yesterday?*

a How many boys went on line for

i less than 60 minutes

ii more than 3 hours

iii did not go on line?

b How many girls went on line for between 1 and 2 hours?

c Boys spend longer on line than girls. True or false?

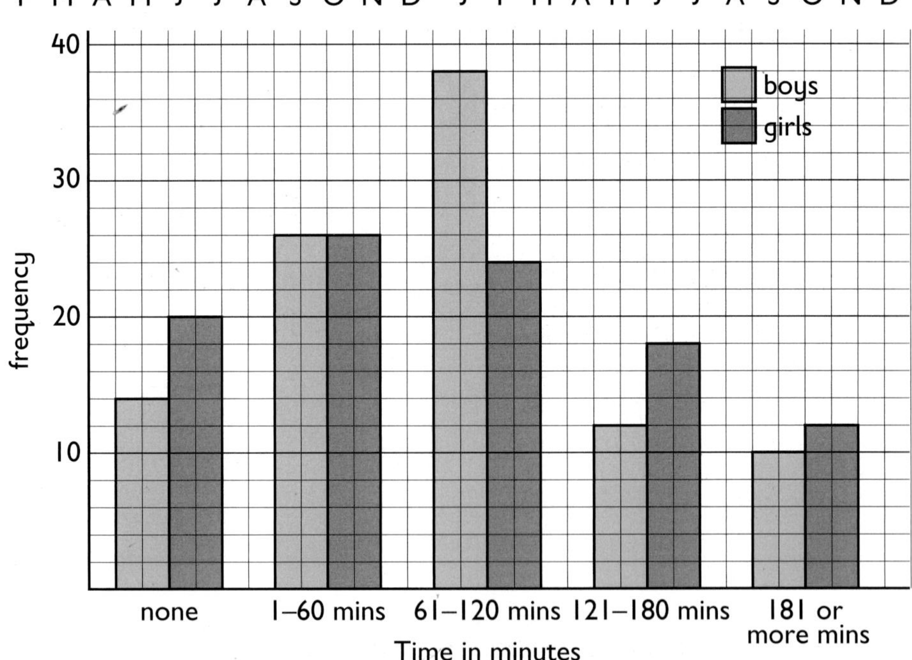

3 Work with a friend. Conduct a survey of pupils in your school.

Ask the question, How long did you spend going on line after school yesterday?

Make a frequency diagram for your data.

Use a suitable computer program to make diagrams, graphs and charts.

Compare your results with the survey in question 2.

Bar-line graphs and compound bar charts

The table shows how space is allocated by some local newspapers.

Title	No of pages	News	Sport	Pictures	Other
Mail	60	20%	50%	25%	5%
Post	40	25%	40%	25%	10%
Herald	48	35%	25%	20%	20%
Times	80	40%	35%	15%	10%
Echo	72	20%	25%	40%	15%

These graphs show the number of pages in each newspaper.

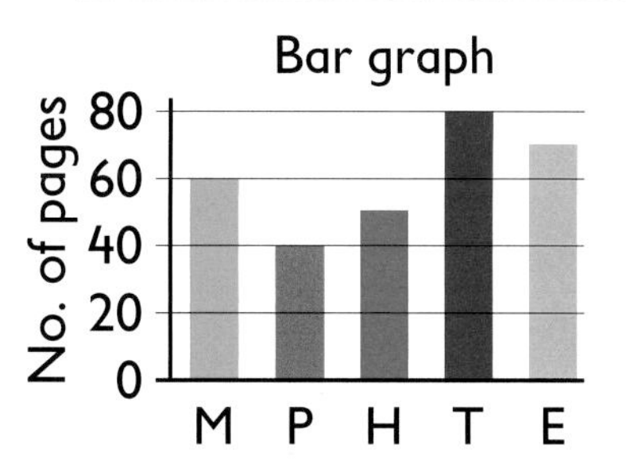

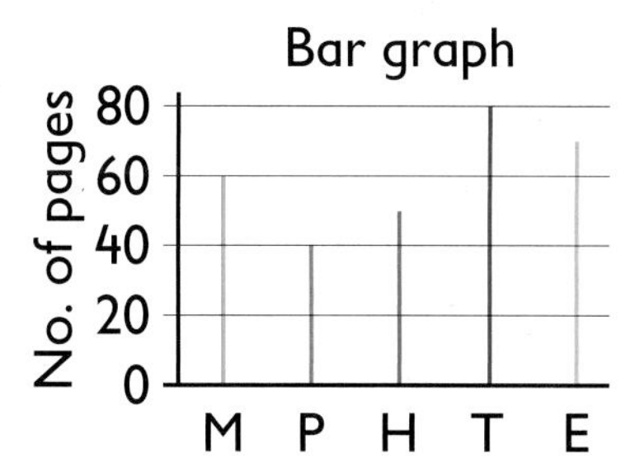

1 Use the information in the table.
Draw bar-line graphs to show how space is allocated by newspaper to:

a News **b** Sport **c** Pictures **d** Other

2 Dan wanted to show how the Mail allocates space to each section.

He drew a compound bar chart.

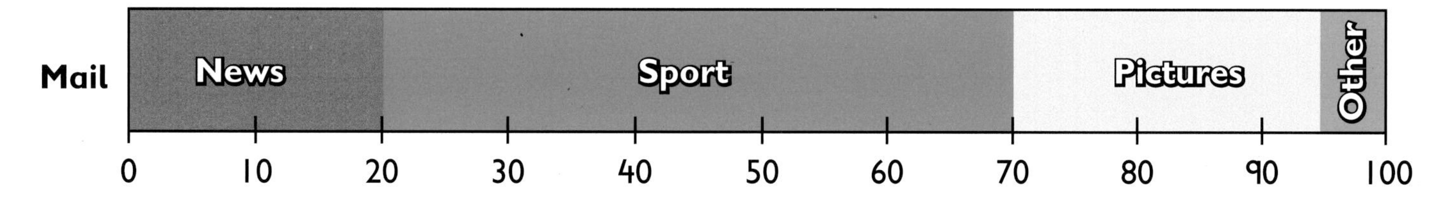

Draw compound bar charts to show how space is allocated by:

a the Post **b** the Herald

c the Times **d** the Echo

> Draw a rectangle, 10 cm by 1 cm.
> Mark the lower side in intervals of 1 cm.
> Label the scale in multiples of 10.

3 For each newspaper, work out the number of pages in the Sports Section.

4 Find which newspaper gives the least space to 'News'.

Pie charts

You need:
- a copy of RCM 49
- colouring materials

1 Kevin kept a record of breakfast orders he took on Friday.

- **a** Calculate the total frequency.
- **b** Convert the frequencies to percentages.
- **c** Draw a percentage pie chart for the table.
- **d** Colour the sectors and make a key.

Meals on Wheels Cafe

Order	Tally	Frequency
Cereal	卌 I	6
Bacon & egg	卌 卌 IIII	
Sausage & egg	卌 卌 II	
Full breakfast	卌 卌 卌 III	

2 For each table:

- **a** Convert frequencies to percentages.
- **b** Draw a percentage pie chart.
- **c** Colour the sectors and make a key.

Type of transport using cafe

Lorry	Bus	Car	Van
54	18	72	36

Sales of hot and cold drinks

Tea	Coffee	Soft drinks	Milk	Fresh orange
250	100	75	50	25

3 Kevin put questionnaires at each table.

Do you consider your health when choosing what to eat?

The table shows how his customers replied.

- **a** Record the information in percentage pie charts.
- **b** Use the table to draw a percentage pie chart for the combined male and female responses.
- **c** Kevin got 200 male and 400 female customer responses.

  Write down 5 facts that you can interpret from your pie charts.

Response	Male	Female
never	15%	12%
sometimes	28%	18%
quite often	37%	46%
very often	15%	17%
always	5%	7%

4 Work with a friend.

Conduct a survey of pupils in your class.

Use a suitable computer program to generate pie charts.

Compare your results with the survey in question 3.

How likely?

1 Write the word that best describes each event.

IMPOSSIBLE | UNLIKELY | CERTAIN | LIKELY

a I will swim in the pool on my holiday.

b I will walk on the moon.

c I will score 6 when I roll the dice.

d The day after Sunday will be Monday.

e Next week will be 8 days long.

f It will rain tomorrow.

g I will have a birthday next year.

h The Queen will visit our school.

i Hens will lay hard-boiled eggs.

j The sun will rise tomorrow.

k Scotland will win the next World Cup.

l I will eat some fruit tomorrow.

2 Mandy has these coins in her purse.

She takes out a coin at random.

How likely is it that the coin is:

a round?

b a heptagon?

c £2?

d silver?

e less than £1?

f more than 20p?

Choose one of the four words in Q 1 to describe the likelihood.

Higher or lower? **An activity for 2 players**

You need:
- 1 to 10 number cards (per pair)
- 30 counters (per pair)

Place the ten cards face down and in a random order.

Turn one card over.

- Players take it in turn to say whether the next card turned over is likely to be higher or lower than the card just turned.
- They give a reason for each response.
- They take a counter if they guess correctly.
- Continue until all the cards have been turned over.
- Play the game two more times.
- The player with the most counters is the winner.

Using a probability scale

1

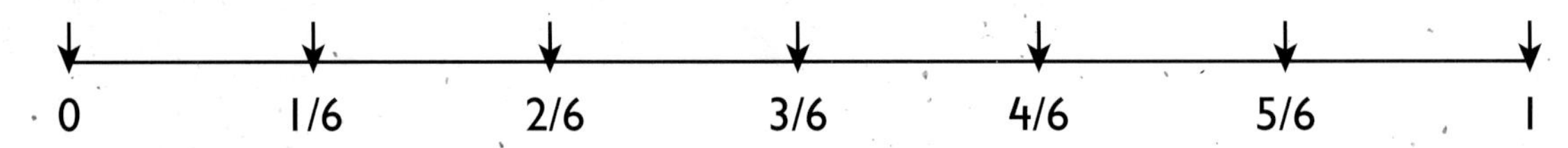

You have a fair 1 to 6 dice. The probability of rolling a 4 is $\frac{1}{6}$ because 4 occurs once out of a total of 6 different possibilities.

Copy and complete the table for these rolls of the dice.

Event	Chance	Probability
a 4	1 in 6	$\frac{1}{6}$
a 2		
an odd number		
zero		
more than 3		
less than 6		
a 3 or a 5		
a number between 0 and 7		

2 Joe chose an apple at random from this display.

a What is the probability that the apple chosen:

i is red?

ii is in an even row?

iii is in the 4th column?

iv is in an odd row and an even column?

b Position the fractions on this probability scale.

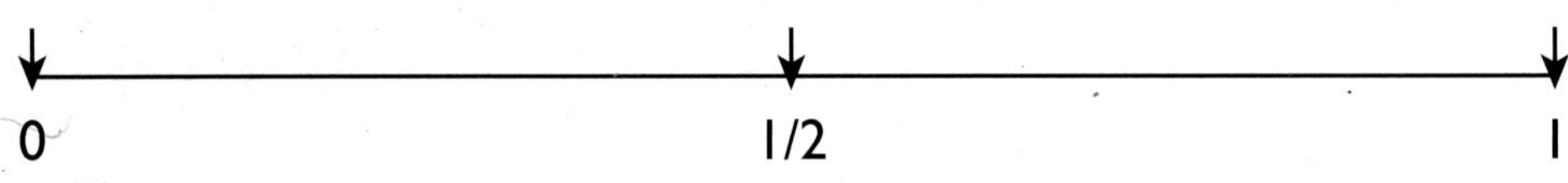

What's the chance? An activity for 2 players

You need:
- a supply of interlocking squares and triangles in red, blue and green (per pair)

a Build 2 different cubes with a 50% chance of landing on a red or blue face.

b Build a cube with a 1 in 3 chance of landing on a red, blue or green face.

c Build a cube with 1 red, 2 blue and 3 green faces.

Draw their position on a probability scale.

d Build an octahedron in 2 colours with

i a 25% chance of landing on red.

ii a $\frac{5}{8}$ chance of landing on blue.

e Build an octahedron in 3 colours with

i a 1 in 8 chance of landing on green.

ii a 75% chance of landing on blue.

Identifying outcomes

I can write this as P(Heads) = $\frac{1}{2}$ P(Tails) = $\frac{1}{2}$

Samir tosses a fair coin.

There are two outcomes: heads or tails

The probability of heads is $\frac{1}{2}$.

The probability of tails is $\frac{1}{2}$.

1 You take a sweet at random from the bag.

- **a** Name the possible outcomes.
- **b** Find the probability of each outcome.

2 There are 5 boxes in a row at the fairground stall.

Two of them contain a prize.

- **a** Name the two outcomes.
- **b** Find the probability of winning.

3 You choose a letter of the alphabet at random.

- **a** Name the two outcomes.
- **b** Find the probability of each outcome.

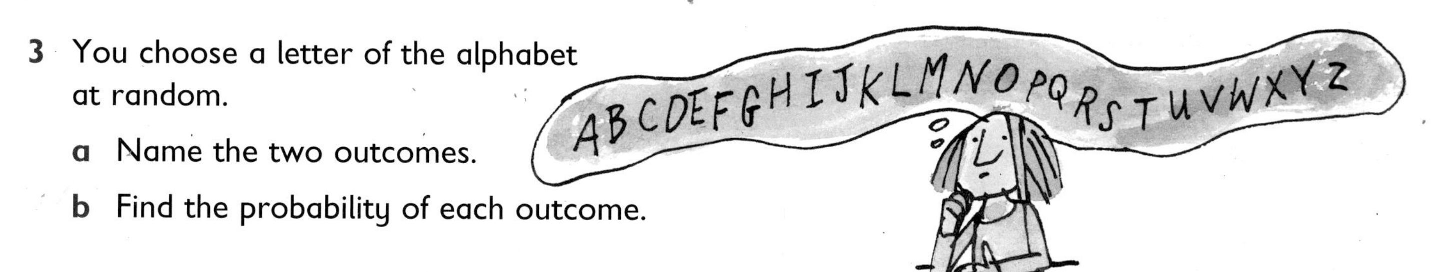

4 You choose a letter at random from this word.

- **a** List the possible outcomes.
- **b** Find the probability of each outcome.
- **c** Do the same for this word.

5 You shuffle these playing cards and place them face down on the table.

- **a** List the possible outcomes.
- **b** Find the probability that the card you turn over is:

i a picture card	**ii** red
iii a diamond	**iv** even
v a spade	**vi** not a club
vii not a picture card	**viii** 4 of hearts

Probability experiments

You need:
- 6 interlocking squares
- Blu-tack
- washable felt tip pen
- 4 cubes (1 of each colour)
- bag

Two experiments to try with a friend.

1 **a** Complete the table for a 1–6 fair dice

Outcome	1	2	3	4	5	6
Probablility						

b Make a net of a cube with 6 interlocking squares. Attach some Blu-tack to one face. Label the faces of the cube 1 to 6.

c Draw a frequency table in your book.

d Take turns of rolling the biased cube. Record each outcome as a tally mark. Do this 50 times.

e Compare your experimental probabilities with the table for a fair dice. Write what you notice.

Outcome	Tally	Frequency
1		
2		
3		
4		
5		
6		

2 **a** Draw this table in your exercise book.

b Put 4 different coloured cubes in a bag.

c Before you take a cube guess its colour. Then take a cube at random from the bag.

d Mark the 1st column ✓ if right or ✗ if wrong. Put the cube on the table.

e Carry on until you have taken out all 4 cubes.

f Repeat the experiment 10 times.

g Find the probability of being right on the 1st guess and on the 4th guess.

Experiment	Guesses			
Number	1st	2nd	3rd	4th
1				
2				
3				
4				

3 Catherine and Jean put 240 identical, but differently coloured counters into a bag. They made a probability scale to show the chance of getting, at random, a counter of a certain colour. How many counters of each colour did they put in the bag?

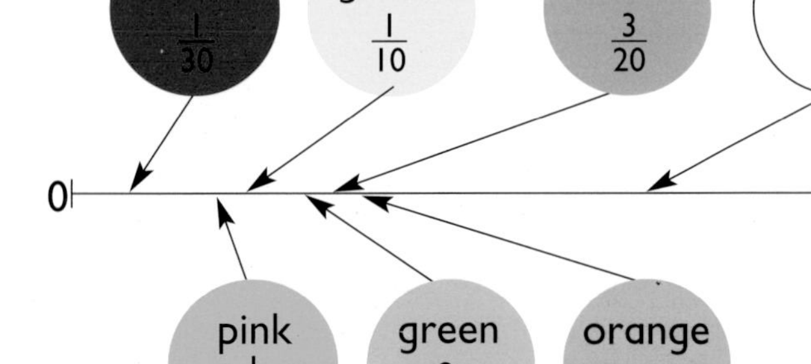

The sum of 2 dice

An activity for 2 children.

You need:

- a red 1–6 dice per pair
- a blue 1–6 dice per pair
- a copy of RCM 50

1 The experiment.

a Work with a partner:

Child A: Roll both dice and find the outcome by adding the 2 scores. Do this 9 times.

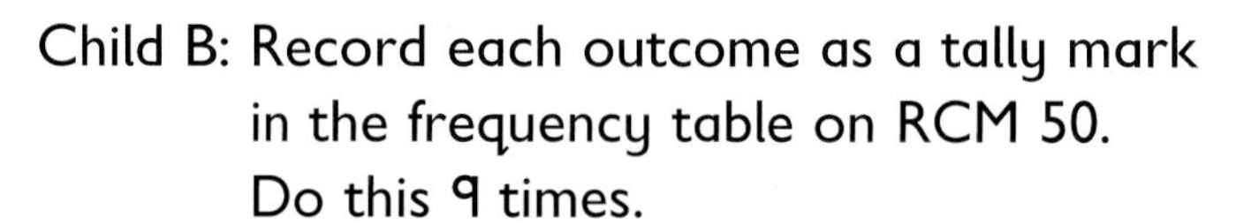

Child B: Record each outcome as a tally mark in the frequency table on RCM 50. Do this 9 times.

b Take 6 turns each of rolling the dice.

c Complete the frequency column in the table. You should have 108 entries altogether.

d Divide each frequency by 3 and complete the column f ÷ 3.

Outcome	Tally	Frequency (f)	f ÷ 3
1	|		
2	|||		
3	||		
4	||||/		
5	|		

2 The theoretical probabilities.

a Record in the table all the possible probabilities for the sum of the scores on a red and a blue dice.

b Copy and complete:
The total of possible probabilities is ___.

red dice

blue dice

+						
		5				

3 The block graph.
Make a block graph to show the theoretical probabilities of question 2.

4 The results.
Look at the last column (f ÷ 3) in the frequency table. Look at the table of probabilities and the block graph for the sum of 2 dice.

Compare your experimental probabilities with the theoretical probabilities. Write what you notice.

5 What if you increase the size of the trial by including the experimental results of other children? Can you predict what will happen to the frequencies?